An Introduction to the Dynamics of El Niño and the Southern Oscillation

An Introduction to the Dynamics of El Niño and the Southern Oscillation

ALLAN J. CLARKE
Department of Oceanography
Florida State University
Tallahassee, FL, USA

ELSEVIER

Amsterdam • Boston • Heidelberg • London
New York • Oxford • Paris • San Diego
San Francisco • Singapore • Sydney • Tokyo

Academic Press is an imprint of Elsevier

Academic Press is an imprint of Elsevier
84 Theobald's Road, London WC1X 8RR, UK
Radarweg 29, PO Box 211, 1000 AE Amsterdam, The Netherlands
Linacre House, Jordan Hill, Oxford OX2 8DP, UK
30 Corporate Drive, Suite 400, Burlington, MA 01803, USA
525 B Street, Suite 1900, San Diego, CA 92101-4495, USA

First edition 2008

ISBN: 978-0-12-088548-0

For information on all Academic Press publications
visit our website at books.elsevier.com

Transferred to Digital Printing 2009.

Working together to grow
libraries in developing countries

www.elsevier.com | www.bookaid.org | www.sabre.org

ELSEVIER   BOOK AID
             International   Sabre Foundation

To my wife, Colette,
and children
Elizabeth, Deborah and Julia.

CONTENTS

5 Wind-Forced Equatorial Wave Theory and the Equatorial Ocean Response to ENSO Wind Forcing   89

6 Sea Surface Temperature, Deep Atmospheric Convection and ENSO Surface Winds   113

PREFACE

Growing up in Australia, the driest inhabited continent, droughts were frequent and long. These droughts are not local to Australia, but rather are linked to large-scale climate anomalies throughout the tropics and a large fraction of the rest of the globe. Concurrent with these atmospheric events are dramatic changes in the ocean: the equatorial Pacific Ocean warms and cools over more than a quarter of the earth's circumference and internal changes in the coastal Pacific and Indian Oceans affect fishing in regions as far apart as California and the western and southern coasts of Australia. All these events are connected to El Niño and the Southern Oscillation (ENSO), the major year-to-year climate signal on earth.

Especially since the early 1980s, thousands of scientists from many countries have realized the importance of this climate phenomenon and have studied it. One major difficulty is that ENSO involves both the ocean and the atmosphere and most scientists who study El Niño and the Southern Oscillation have been trained solely as either oceanographers or meteorologists. The purpose of this book is to overcome this difficulty by providing an introduction to this coupled ocean–atmosphere phenomenon.

Chapters 2–7 of the book are based on an upper level graduate course, Equatorial Dynamics, which I have taught for many years in the Oceanography Department at Florida State University. I am indebted to the many talented men and women who have taken the course and have made my professional life so enjoyable. I am especially indebted to one of them, David Battisti, for suggesting several years ago that I write this book.

The book begins with a history of ENSO in Chapter 1 and an overview of the key ENSO observations in Chapter 2. These chapters contain no mathematics. The key ocean/atmosphere interaction occurring during El Niño is in the equatorial Pacific, so to understand El Niño it is important to understand the various types of equatorial ocean waves for these can carry energy from one side of the Pacific to the other. Assuming a knowledge of the equations of fluid mechanics on a rotating sphere, Chapter 3 shows how these ocean waves arise and describes those relevant to the year-to-year fluctuations associated with El Niño. These waves propagate and reflect from both the solid eastern Pacific Ocean boundary and the gappy western Pacific boundary, and Chapter 4 describes how these reflections occur. Such reflections are not just of theoretical interest – they profoundly affect not only ENSO dynamics but also marine ecosystems along thousands of kilometers of coastline of North, Central and South America as well as the western and southern coasts of Australia. ENSO's effect on marine life is discussed in Chapter 12. Chapter 5 introduces the wind forcing and shows how the ocean response to that forcing can be described in terms of wind-forced long ocean waves. After this, the observed year-to-year equatorial ocean response to wind forcing is presented and explained theoretically.

Having discussed how the atmosphere drives the ocean in Chapter 5, the next step, in Chapter 6, is to discuss how the ocean drives the atmosphere through the influence of the sea surface temperature. A crucial part of the atmospheric response to the sea surface temperature is the equatorial zonal surface wind, for the ocean is very sensitive to this at the very low ENSO frequencies of interest. In Chapter 6, this equatorial surface wind is understood using the equatorial wave concepts introduced in Chapter 3.

Now that the uncoupled atmosphere drives ocean and ocean drives atmosphere problems have been considered, Chapter 7 discusses coupled ocean–atmosphere models of ENSO. The emphasis in this chapter is on the physics of the two major ENSO paradigms and a description of a widely used ENSO model that was the first dynamical model to provide an El Niño forecast.

ENSO events tend to have their greatest amplitude near the end of the calendar year, i.e., they tend to be phase-locked to the seasonal cycle. This curious behavior is discussed in Chapter 8, and then, in Chapter 9, basic theory for upper air response and climate signal linkages far from the equatorial Pacific is presented. Chapters 10 and 11 describe prediction of the ENSO signal using dynamical and statistical models and attempt to compare their performance.

The book focuses on introducing the basic dynamics of ENSO and consequently many ENSO topics have not been covered. These include decadal and lower frequency climate variability and their influence on ENSO, the possible influence of anthropogenic effects on ENSO, ENSO variability thousands and hundreds of thousands of years ago, coral, tree-ring, ice core and other proxy ENSO indicators and the applications of ENSO to agriculture, hurricane, drought and wildfire prediction, vector-borne diseases and stream flow. Chapter 12 on ENSO's influence on marine and bird life has been included both because of my own research interests and also because of the close connection, in several sections of Chapter 12, to theory in prior chapters. Notwithstanding these connections, most of Chapter 12 can be understood by readers without a strong mathematical background.

In this book I have tried to use algebraic symbols that are as standard and simple as possible. In a very few cases the same symbol can mean different things but in these cases the meaning is clear from the context.

As is true for most scientists, it bothers me when I do not understand something. Consequently, I have tried to explain the relevant

El Niño/Southern Oscillation dynamics as simply as possible. This challenge sometimes resulted in published research during the several years I wrote this book, and as a result, this volume contains many more citations to my own work than would normally be expected. Please forgive me for this distortion.

I am indebted to many colleagues for their help. Jim O'Brien, major professor for many graduate students at Florida State, provided many of the students who have taken my Equatorial Dynamics course. Several colleagues read parts of the manuscript. Grant Branstator commented on part of Chapter 9, Tony Barnston, Huug van den Dool, Cecile Penland and John Knaff commented on parts of Chapter 11 and Ken Brink likewise for Chapter 12. Others generously provided data to construct figures or provided electronic versions of them: Diane Stokes and Dick Reynolds (Fig. 2.4); Sim Larkin and Ed Harrison (Figures 8.3–8.5); Kwang-Yul Kim (Fig. 9.3); Eric DeWeaver (Figs 9.4 and 9.5); Dake Chen, Ben Kirtman, Jong-Seong Kug, Geert Jan van Oldenborgh and Suru Saha (Fig. 10.1); Dake Chen (Fig. 10.2); John Knaff, Cecile Penland and Huug van den Dool (Fig. 11.1); Chris Chubb (Fig. 12.10); and Patrick Lehodey (Fig. 12.13).

I am especially grateful to two people at Florida State. Steve Van Gorder plotted several figures, read and provided helpful constructive criticism on the manuscript, and Paula Jahromi skillfully drafted the figures and typed the text. It is a pleasure working with Steve and Paula.

Finally, I am thankful to my wife, Colette, and children, Elizabeth, Deborah and Julia, for their love and encouragement and for allowing me the extra time beyond normal working hours to complete this book. Deborah has a degree in art and I thoroughly enjoyed working with her on the design for the book's cover.

Allan J. Clarke
July 2, 2007.

CHAPTER ONE

INTRODUCTION

Broad beaches, bare hills and deserts form the landscape of much of the Peruvian coast. But while the land is barren, the adjacent sea is luxuriant and for centuries has been harvested by local fishermen. For most of the year the coastal water is cold and the local surface current is northward along the coast. But beginning around Christmas time, the current flows southward and the water warms. At least since the late nineteenth century, fishermen from the port of Paita in northern Peru have called this warm current the current of 'El Niño' (the Child Jesus) because of its appearance around Christmas time.

Occasionally the southward flow and warm water lasts longer than a few months, well into the next year. In 1891 the warm El Niño current was unusually strong and the normally desert coast was soaked by rain. A few weeks later pasture covered the desert. This inspired Carranza (1891), Carillo (1892) and Eguiguren (1894) to document the current and to suggest, probably for the first time, a link between the current and the change in the regional climate, especially the Peruvian coastal rainfall.

About a decade earlier on the other side of the world in colonial India the Indian monsoon of 1877 failed, resulting in catastrophic drought and famine. This year was also a year when the Peruvian desert blossomed, but it is only recently that we have understood that these events are related. We now know that the coastal sea surface temperature (SST) changes associated with the El Niño current not only occur near the Peruvian coast, but also extend along the equator for a distance of about one quarter of the way around the world! These temperature changes are caused by changes in the equatorial Pacific westerly winds which are in turn related to a huge seesaw in surface atmospheric pressure that rocks slowly and irregularly every 2–5 years; one end of the seesaw is in the tropical western Pacific, Australia, India and the eastern Indian Ocean while the other is in the eastern Pacific. The pressure is a seesaw in the sense that when the western Pacific, Australia, India and Indian Ocean have anomalously high surface pressure, the eastern Pacific has anomalously low surface pressure and vice versa.

Henry Blanford, the first Imperial Meteorological Reporter to the Government of India, first noticed the Indo-Australian end of the seesaw when he reported that the 1877 failure of the Indian monsoon was associated with extremely high surface pressure values over much of the seesaw's

1

western end. Hildebrandsson (1897) was probably the first to detect each end of the seesaw when he reported that the surface pressure in Sydney, Australia, and Buenos Aires, Argentina, tended to be out of phase. Further analysis of several stations worldwide by Sir Norman Lockyer and his son William (Lockyer and Lockyer 1904) and Frank Bigelow (1903) from the United States then revealed the existence of the seesaw.

Sir Gilbert Walker, the second Director-General of Observatories in India from 1904–1924, named the seesaw the 'Southern Oscillation' (Walker 1924). He used this name because a synthesiↄ of statistical correlations between stations all over the world suggested to him that there were three coherent large-scale climatic oscillations – the 'Southern Oscillation' involving the seesaw of atmospheric pressure and rainfall across the Indo-Pacific Region and two northern oscillations, one associated with opposite phase pressure between the Azores and Iceland (the North Atlantic Oscillation) and the other opposite phase pressure between Alaska and the Hawaiian Islands (the North Pacific Oscillation).

Analysis of the data by Walker (1923, 1924) and Walker and Bliss (1932, 1937) revealed a strange property of the Southern Oscillation, namely, that lag correlations are high for much of the year but fall sharply across the boreal spring, i.e., there is a boreal spring persistence barrier. This means, for example, that if a seasonal June–July–August (JJA) value of the Southern Oscillation is (say) positive, it will tend to stay positive over the next several months but a December–January–February (DJF) value could remain positive or become negative across the boreal spring. Thus long-range persistence predictions are possible from JJA but not from DJF. This property, fundamental to prediction and an understanding of the Southern Oscillation, has been the subject of much recent research and will be discussed in Chapter 8.

Walker's examination of the world data he collected enabled him to establish that the Southern Oscillation is correlated with Indian summer rainfall, northeast Australian, South African and South American (Southern Hemisphere) summer rainfall, Chile winter rainfall, Nile floods, southwest Canada winter temperature, Java December–February rainfall and air temperature in New Zealand and the tropics in various places. He set up simple formulae, based on these data, for making long-range forecasts. These long-range forecasts were not forecasts of daily weather, but rather average weather over (say) a season of the year. It is in this sense that we make long-range forecasts today; we do not try to predict (say) the temperature in Tallahassee, FL, on Christmas Day 2007 but rather the average winter temperature for 2007–2008.

There was acceptance but also some criticism of Walker's work during his lifetime. Probably the most negative was his obituary in the 1959

Quarterly Journal of the Royal Meteorological Society (p. 186). The obituary's author, in describing Walker's correlation studies, stated 'Walker's hope was presumably not only to unearth relations useful for forecasting but to discover sufficient and sufficiently important relations to provide a productive starting point for a theory of world weather. It hardly seems to be working out like that.' But in fact it did work out like that – many of Walker's statistical relationships were verified with longer records and these relationships are today foundational to the theory of climate variability.

When Walker was working on the statistical correlations between atmospheric pressure and rainfall on a global scale, Brooks and Braby (1921) reported key observations of rainfall and wind at Pacific Islands near the International Date Line in the west-central equatorial Pacific. They found that long dry and wet spells were coherent over a huge equatorial area stretching from 170°E to 155°W. The dry spells were associated with easterly winds and the wet spells with wind convergence, frequent calm conditions and both easterly and westerly winds. We now know that this equatorial region near the Date Line is directly associated with the generation of the Southern Oscillation and El Niño (see Chapter 2).

Probably Leighly (1933) was the first to propose that SST anomalies could influence rainfall in the central equatorial Pacific. He also suggested that westward equatorial surface pressure gradient should be associated with easterly winds, low water temperature and drought. His paper was ignored, however, and it was not until around the time of Walker's death in 1958 that El Niño and the Southern Oscillation were linked. Berlage and de Boer (1960) found that SST at the Peruvian port of Puerto Chicama was strongly correlated with the Southern Oscillation. Berlage (1957, 1961) even proposed a theoretical explanation for El Niño based on remote wind forcing and also suggested a coupled air–sea interaction mechanism for the Southern Oscillation. Although his theoretical ideas were incorrect, the idea that El Niño and the Southern Oscillation are linked is foundational to the modern theory of El Niño and the Southern Oscillation (ENSO).

In 1969, Professor Jacob Bjerknes outlined a coupling mechanism between the equatorial Pacific Ocean and atmosphere that seemed more reasonable (Bjerknes 1969). He noted that the equatorial Pacific surface water was cold in the east and warm in the west and that this could drive a circulation in a vertical equatorial plane. In this circulation relatively cold, dry air sinks over the cold water in the eastern Pacific and then moves eastward along the equator to the warm equatorial Pacific where it is heated and supplied with moisture. It then rises over the warm water and returns aloft to the eastern equatorial Pacific, thus completing the circulation. Bjerknes named this equatorial circulation the 'Walker Circulation' after Sir Gilbert Walker.

Bjerknes also reasoned that an increase in the Walker Circulation caused a positive feedback in the following sense. If the Walker circulation increases then the surface equatorial easterly wind increases, causing increased upwelling of cold water in the eastern Pacific (see Chapter 2 for a discussion of this). This increases the contrast between the warm water in the west and the colder water in the east, increasing the sinking air in the east and leading to a stronger Walker Circulation. Similarly, a decrease in strength of the Walker Circulation decreases the upwelling in the east, making the water warmer there and hence reducing the sinking air and the strength of the Walker Circulation. Thus Bjerknes identified a positive air–sea feedback mechanism which would explain trends of increasing or decreasing strength in the Walker Circulation. But the turnabout between alternating trends, which is needed to complete a physical explanation for a climate oscillation, was to him 'not yet quite clear.'

In fact it would take nearly 20 years before the 'turnabout' began to be understood. What was needed was a coupled ocean–atmosphere model, i.e., a model consisting of separate ocean and atmosphere models linked together so that each influences the other. Specifically, winds from the atmosphere model should drive the ocean model which should then change its SST and hence change the heating of the atmosphere. Changed atmospheric heating should then change the model wind which in turn should change the model ocean SST, etc.

But in 1969, when Bjerknes stated the 'turnabout' problem, little was understood of how the equatorial ocean responds to wind forcing and how the tropical atmosphere responds to heating. Key to the ocean problem is a knowledge of equatorial ocean waves, for these can carry energy thousands of kilometers along the equator from where they are forced. The first to discover equatorial waves was Gordon Groves (1955) when he was a PhD student at the Scripps Institution of Oceanography at the University of California. He had developed equatorial wave theory to try to understand sea level fluctuations at Canton Island in the equatorial Pacific. However, other than in his dissertation, Groves never published his equatorial wave results. This theory was therefore essentially unknown until Japanese meteorologist Taroh Matsuno (1966) published it more than a decade later. Cambridge professor Sir James Lighthill (1969) laid the ground work for how the equatorial ocean would respond to the wind at the large time and spatial scales relevant to El Niño, but he was apparently unaware of Matsuno's work, for he did not include one of the waves, the equatorial Kelvin wave, in his otherwise elegant theory. Five years later another Cambridge researcher, Adrian Gill, generalized Lighthill's theory to include this wave (Gill and Clarke 1974). The equatorial Kelvin wave is of considerable importance because it is the only wave, at ocean basin scales, that can carry energy eastward. Oceanographer Stuart Godfrey (1975) at the Commonwealth Scientific and Industrial Research Organization (CSIRO) in Sydney, Australia, suggested that El Niño could result when equatorial trade winds weakened and an equatorial Kelvin wave propagated eastward to the South American coast. An extensive study of sea level, ocean temperature and wind by Professor Klaus Wyrtki (1975) at the University of Hawaii provided observational support for this idea. By now several in the oceanography community realized the importance of El Niño and several papers on El Niño and equatorial ocean dynamics soon followed (e.g., Cane and Sarachik 1976, 1977, 1979, 1981; Hurlburt et al. 1976; McCreary 1976; Moore and Philander 1977).

So by the late 1970s enough was understood about ocean dynamics to construct the ocean part of the coupled problem. The atmospheric part was also available from the simple model of Matsuno (1966). Fourteen years later Adrian Gill (1980) described a model that was essentially the same as Matsuno's, but Gill obtained simple model solutions and interpreted them physically for quite general atmospheric heating. The Matsuno/Gill model has since become widely used in ENSO studies.

From the late 1970s oceanographers and meteorologists began to think about coupled ocean–atmosphere ENSO dynamics. Mark Cane, at the time a professor at the Massachusetts Institute of Technology, was inspired by Bjerknes's work. Building on the forced ocean wave ideas of the 1970s, he built an efficient numerical ocean model (Cane and Patton 1984) and then, with Stephen Zebiak, coupled a Matsuno/Gill atmospheric model to it. This was one of the first coupled ENSO models and the first dynamical model (Cane et al. 1986) to be used to predict El Niño. It is still (with some modification) used to forecast El Niño to this day. The model is described in more detail in Sections 7.4, 10.2, 10.3 and 10.5.

Shortly thereafter, NASA researcher Paul Schopf (1987) suggested that the ENSO 'turnabout' might be due to ocean waves providing a delayed negative feedback. He called the resultant theory 'delayed oscillator theory' and, with fellow NASA researcher Max Suarez (Schopf and Suarez 1988; Suarez and Schopf 1988), showed how it worked with a numerical model and with a simpler theory to illustrate the essential physics. In his PhD thesis David Battisti, a student at the University of Washington, Seattle, showed that delayed oscillator physics explained the dynamics of the Cane and Zebiak model (Battisti 1988) and with Tony Hirst (Battisti and Hirst 1989) slightly modified the Suarez and Schopf delayed oscillator theory. On the other hand, researchers Stephen Zebiak and Mark Cane (Zebiak and Cane 1987) suggested, based on results of their coupled ocean–atmosphere model, that the volume of warm water in the upper equatorial Pacific Ocean is a 'critical element' in the dynamics. Later theory and observations by Fei-Fei Jin (1997), Christopher Meinen and Michael McPhaden (Meinen and McPhaden 2000) supported a 'recharge oscillator' theory for ENSO turnabout. These theoretical ideas will be discussed in more detail in Chapter 7.

Bjerknes's original ideas about the Walker Circulation depend on the SST contrast between the eastern and western equatorial Pacific. Consequently they suggest that *eastern* equatorial Pacific SST anomalies should have a profound influence on ENSO dynamics, particularly since these SST anomalies are bigger than those in the west. But a careful analysis of the 1982–1983 El Niño by Adrian Gill and meteorologist Gene Rasmusson (1983) showed that the most important ENSO coupling between ocean and atmosphere occurred not in the eastern equatorial Pacific, but rather at the edge of a huge pool of warm water usually located 10 000 km away in the western equatorial Pacific. The atmosphere over the warm pool is strongly heated because there the warm moist air rises and the water vapor condenses and releases latent heat. Gill and Rasmusson found that as this warm water moves eastward, the deep atmospheric convection and heating moves with it. This atmospheric heating generates westerly (eastward) winds (for details, see Section 6.8) which force the equatorial ocean flow eastward and cause the warm pool to be moved even further eastward. By this process the warm pool should be moved right across the equatorial Pacific, as indeed it was in 1982 and, more recently, in 1997. Later analysis by Fu et al. (1986) showed that normally the eastward displacement of the warm pool during El Niño is typically smaller (about 2000 km) and that the eastward and westward warm pool displacements along the equator are intimately associated with the Southern Oscillation and El Niño (eastward displacement) and La Niña (westward displacement). The term La Niña (Spanish for the girl) refers to the opposite phase of El Niño when the eastern equatorial Pacific is colder rather than warmer than usual. Analysis of satellite sea level observations by French researchers Joel Picaut and Thierry Delcroix confirmed the suggestion by Adrian Gill (1983) that the east–west displacements were mainly due to currents associated with equatorial Kelvin and Rossby waves. Warm pool displacement theory will be discussed in more detail in Chapter 7.

A thorough analysis of SST, wind fields and precipitation patterns by Gene Rasmusson and Thomas Carpenter (1982) revealed a paradoxical property of El Niño and the Southern Oscillation, namely, that they are locked in phase with the seasonal cycle. Phase-locking to the seasonal cycle is paradoxical because El Niño and the Southern Oscillation are oceanic and atmospheric climate *anomalies*, i.e., variability that has had the average seasonal cycle removed. In Chapter 8 we will discuss this phase-locking and its relationship to the boreal spring persistence barrier discovered by Sir Gilbert Walker.

Research interest in ENSO has blossomed since the beginning of the 1980s. Then there were just a few researchers, limited measurements and poor communication. Indeed, in October 1982 when scientists met in Princeton, NJ, to discuss plans for a program to study El Niño, the biggest El Niño of the century was occurring and no one there was aware of it! Now up-to-date estimates of winds, SST, upper ocean temperature, sea level, rainfall, atmospheric temperature and other relevant parameters are readily available from satellites and in situ measurements. Much of the in situ data comes from the Tropical Atmosphere Ocean (TAO) array begun by Stan Hayes of the Pacific Marine Environmental Laboratory in Seattle, United States. The array, now supported by an international consortium, consists of about 70 moorings spanning the tropical Pacific. The moorings telemeter oceanographic and meteorological data in real time via the Argos satellite system. Hundreds of scientists from many countries use these and other available data to predict and understand El Niño. A major step forward in the improvement and use of seasonal to interannual forecasts for the benefit of society was taken in the United States in late 1996 with the formation of the International Research Institute for Climate Prediction.

In the 1950s we began to make weather forecasts and use them. Now we have begun to make long-range climate forecasts and apply them. It's an exciting time.

▶ **REFERENCES**

Battisti, D. S., 1988: Dynamics and thermodynamics of a warming event in a coupled tropical atmosphere–ocean model . *J. Atmos. Sci.*, **45**, 2889–2919.

Battisti, D. S., and A. C. Hirst, 1989: Interannual variability in a tropical atmosphere–ocean model: Influence of the basic state, ocean geometry and nonlinearity. *J. Atmos. Sci.*, **46**, 1687–1712.

Berlage, Jr, H. P., 1957: *Fluctuations of the general atmospheric circulation of more than one year, their nature and prognostic value.* Mededlingen en Verhandelingen No. 69, Koninklijk Meteorologische Instituut, Staatsdrukkerijs-Gravenhage, Netherlands, 152 pp.

Berlage, Jr, H. P., 1961: Variations in the general atmospheric and hydrospheric circulation of periods of a few years duration affected by variations of solar activity. *Ann. N.Y. Acad. Sci.*, **95**, 354–367.

Berlage, Jr, H. P., and H. J. de Boer, 1960: On the Southern Oscillation, its way of operation and how it affects pressure patterns in the higher latitudes. *Pure Appl. Geophys.*, **46**, 329–351.

Bigelow, F. H., 1903: Studies on the circulation of the atmospheres of the Sun and of the Earth. II. Synchronism of the variations of the solar prominence with the terrestrial barometric pressures and the temperatures. *Mon. Weather Rev.*, **32**, 509–516.

Bjerknes, J., 1969: Atmospheric teleconnections from the equatorial Pacific. *Mon. Weather Rev.*, **97**, 163–172.

Brooks, C. E. P., and H. W. Braby, 1921: The clash of the trades in the Pacific. *Q. J. R. Meteorol. Soc.*, **47**, 1–13.

Cane, M. A., and R. J. Patton, 1984: A numerical model for low-frequency equatorial dynamics. *J. Phys. Oceanogr.*, **14**, 1853–1863.

Cane, M. A., and E. S. Sarachik, 1976: Forced baroclinic ocean motions. I. The linear equatorial unbounded case. *J. Mar. Res.*, **34**(4), 629–665.

Cane, M. A., and E. S. Sarachik, 1977: Forced baroclinic ocean motions. II. The linear equatorial bounded case. *J. Mar. Res.*, **35(2)**, 395–432.

Cane, M. A., and E. S. Sarachik, 1979: Forced baroclinic ocean motions. III. The linear equatorial basin case. *J. Mar. Res.*, **37**, 355–398.

Cane, M. A., and E. S. Sarachik, 1981: The response of a linear baroclinic equatorial ocean to periodic forcing. *J. Mar. Res.*, **39**, 651–693.

Cane, M. A., S. E. Zebiak, and S. C. Dolan, 1986: Experimental forecasts of El Niño. *Nature*, **321**, 827–832.

Carillo, C. N., 1892: Disertacion sobre las Corrientes Oceanicas y Estudios de la Corriente Peruana o de Humboldt. Microficha. *Bol. Soc. Geogr. Lima*, **11**, 52–110.

Carranza, L., 1891: Contra-corriente maritima observada Piata y Pacasmayo. *Bol. Soc. Geogr. Lima*, **1**(9), 344–345.

Eguiguren, V., 1894: Las Lluvias en Piura. *Bol. Soc. Geogr. Lima*, **IV**, 4–20.

Fu, C., H. F. Diaz, and J. O. Fletcher, 1986: Characteristics of the response of sea surface temperature in the central Pacific associated with warm episodes of the Southern Oscillation. *Mon. Weather Rev.*, **114**, 1716–1738.

Gill, A. E., 1980: Some simple solutions for heat-induced tropical circulation. *Q. J. R. Meteorol. Soc.*, **106**, 447–462.

Gill, A. E., 1983: An estimation of sea level and surface-current anomalies during the 1972 El Niño and consequent thermal effects. *J. Phys. Oceanogr.*, **13**, 586–606.

Gill, A. E., and A. J. Clarke, 1974: Wind-induced upwelling, coastal currents and sea-level changes. *Deep-Sea Res.*, **21**, 325–345.

Gill, A. E., and E. M. Rasmusson, 1983: The 1982–83 climate anomaly in the equatorial Pacific. *Nature* (London), **306**, 229–234.

Godfrey, J. S., 1975: On ocean spin-down. I. A linear experiment. *J. Phys. Oceanogr.*, **5**, 399–409.

Groves, G. W., 1955: Day to day variation of sea level. PhD Thesis, Scripps Institution of Oceanography, University of California, Los Angeles, 88 pp.

Hildebrandsson, H. H., 1897: Quelques recherches sur les centres d'action de l'atmosphere. *Kunglica Svenska Vetenskaps-akademiens Handlingar*, **29**, 36 pp.

Hurlburt, H. E., J. C. Kindle, and J. J. O'Brien, 1976: A numerical simulation of the onset of El Niño. *J. Phys. Oceanogr.*, **6**, 621–631.

Jin, F. F., 1997: An equatorial ocean recharge paradigm for ENSO. Part I: Conceptual model. *J. Atmos. Sci.*, **54**, 811–829.

Leighly, J. B., 1933: Marquesan meteorology. *Univ. Calif. Publ. Geogr.*, **6**(4), 147–172.

Lighthill, M. J., 1969: Dynamic response of the Indian Ocean to the onset of the Southwest Monsoon. *Philos. Trans. R. Soc. Lond. Ser. A*, **265**, 45–93.

Lockyer, N., and W. J. S. Lockyer, 1904: The behavior of the short-period atmospheric pressure variation over the earth's surface. *Proc. R. Soc. Lond.*, **73**, 457–470.

Matsuno, T., 1966: Quasi-geostrophic motions in the equatorial areas. *J. Meteorol. Soc. Jpn. Ser. II*, **44**, 25–43.

McCreary, J. P., 1976: Eastern tropical ocean response to changing wind systems, with application to El Niño. *J. Phys. Oceanogr.*, **6**, 632–645.

Meinen, C. S., and M. J. McPhaden, 2000: Observations of warm water volume changes in the equatorial Pacific and their relationship to El Niño and La Niña. *J. Climate*, **13**, 3551–3559.

Moore, D. W. and S. G. H. Philander, 1977: Modeling the equatorial oceanic circulation. In: *The Sea*, Vol. VI, Wiley Interscience, New York, 319–361.

Rasmusson, E. M., and T. H. Carpenter, 1982: Variations in sea surface temperature and surface wind fields associated with the Southern Oscillation/El Niño. *Mon. Weather Rev.*, **110**, 354–384.

Schopf, P. S., 1987: Coupled dynamics of the tropical ocean–atmosphere system. *U.S. TOGA Workshop on the Dynamics of the Equatorial Oceans*, Honolulu, Hawaii, TOGA, 279–286.

Schopf, P. S., and M. J. Suarez, 1988: Vacillations in a coupled ocean–atmosphere model. *J. Atmos. Sci.*, **45**, 549–567.

Suarez, M. J., and P. S. Schopf, 1988: A delayed action oscillator for ENSO. *J. Atmos. Sci.*, **45**, 3283–3287.

Walker, G. T., 1923: Correlation in seasonal variations of weather, VIII. A preliminary study of world-weather. *Mem. Indian Meteorol. Dept*, **24**, 75–131.

Walker, G. T., 1924: Correlation in seasonal variations in weather IX: A further study of world-weather. *Mem. Indian Meteorol. Dept*, **24**, 275–332.

Walker, G. T., and E. W. Bliss, 1932: World weather V. *Mem. R. Meteorol. Soc.*, **4**, 53–84.

Walker, G. T., and E. W. Bliss, 1937: World weather VI. *Mem. R. Meteorol. Soc.*, **4**, 119–139.

Wyrtki, K., 1975: El Niño – The dynamic response of the equatorial Pacific Ocean to atmospheric forcing. *J. Phys. Oceanogr.*, **5**, 572–584.

Zebiak, S. E., and M. A. Cane, 1987: A model El Niño–Southern Oscillation. *Mon. Weather Rev.*, **115**, 2262–2278.

CHAPTER TWO

ENSO in the Tropical Pacific

Contents

2.1. Overview

This chapter contains a broad-brush overview of the most important ENSO observations. We will focus on ENSO in the equatorial Pacific, for this is where ENSO is generated. Remotely forced ENSO variability outside of the equatorial Pacific will be examined in Chapter 9. We will present the observations first (Sections 2.2 and 2.3) and follow these with a physical explanation of El Niño (Section 2.4).

2.2. Spatial Patterns of ENSO in the Tropical Pacific

As noted in the Introduction, during an El Niño, the surface waters of the eastern equatorial Pacific are warmer than usual. There are several El Niño indices based on this observation. They often are defined by an average sea surface temperature (SST) over some part of the eastern equatorial Pacific for a month or several months minus the normal SST value for that month or a period of several months. Figure 2.1 shows two such indices of anomalous

11

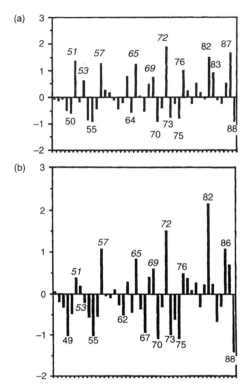

Fig. 2.1 Time series of anomalous SST in degrees Celsius in the equatorial Pacific [6°S–6°N, 180°–80°W] (a) during July–November and (b) December–February. Anomalies are relative to a 1946–1985 mean. Labeled years correspond to prominent warm and cold episodes. Year labels in (b) correspond to the December months: e.g., '57' denotes the season December 1957–February 1958. (Redrawn from Deser and Wallace 1990.)

SST. Notice that the SST anomaly has zero mean and so both warmer than normal conditions (El Niño) and colder than normal conditions (commonly called La Niña which is Spanish for 'the girl') occur irregularly every few years. The typical amplitude for these indices is about 1°C.

Deser and Wallace (1990) have used the above SST anomaly indices to estimate the typical spatial patterns of ENSO by correlating the index with key physical variables in ENSO. The regression coefficient maps of these key variables on the SST index in Fig. 2.1(a) are shown in Fig. 2.2. Multiplying these regressions by 1.4°C yields anomaly fields representative of the five strongest warm tropical Pacific episodes during the period 1946–1985, while multiplying them by −0.8 yields fields representative of the five coldest episodes during the same period.

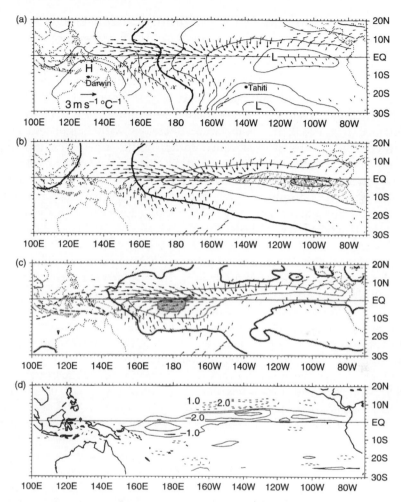

Fig. 2.2 (a) Regressions of July–November sea level pressure and surface wind upon the Fig. 2.1(a) equatorial SST index (defined as July–November SST averaged over 6°N–6°S, 180°W to 80°W), based on the period 1946–1985. Contour interval = 0.25 mb per °C of the SST index; the zero contour is darkened. Wind vectors are shown only for those grid points whose u or v correlations with the SST index exceed 0.4 in absolute value; (b) as in (a) but for SST and surface wind. Contour interval = 0.5°C of the SST index; the zero contour is darkened and regression coefficient >1°C per °C is shaded; (c) as in (a) but for Outgoing Longwave Radiation (OLR) and surface wind. OLR regressions are based on the period 1974–1989 (1978 missing). Contour interval = 10 W m^{-2} per °C of the SST index; the zero contour is darkened and the positive contour is dashed. Values < -20 W m^{-2} per °C are shaded. (d) Divergence of surface wind regressions shown in (a) in units of 10^{-6} s^{-1} °C^{-1}. Solid (dashed) contours indicate anomalous wind convergence (divergence) during warm episodes. (Redrawn from Deser and Wallace 1990).

2.2.1. The western equatorial Pacific region 150°E–150°W where ocean–atmosphere coupling is strongest

During ENSO, large-scale movements of the atmosphere are driven mainly by the anomalous heating which occurs when water vapor condenses to form clouds and rain and latent heat is released. In the troposphere, the lower layer of the atmosphere where almost all water vapor and clouds occur, atmospheric temperature decreases with height and cloud tops therefore tend to be cold. Since colder bodies radiate less heat than warmer ones, clouds are 'seen' by a satellite as regions of low outgoing longwave radiation (OLR). The lowest OLR in the tropics coincides with regions of deep atmospheric convection where storm clouds reach the top of the troposphere. Thus, the anomalously low OLR in the central equatorial Pacific between about 150°E and 150°W in Fig. 2.2(c) shows that this region is more cloudy than normal during a warm ENSO episode and less cloudy than normal during a cold ENSO episode. This figure therefore shows that the main anomalous heating driving the atmosphere during ENSO occurs in this west/central equatorial Pacific region.

Figure 2.2(c) also shows that during a warm ENSO episode, surface winds are westerly under and to the west of the west/central equatorial region of strong ENSO atmospheric heating. Such a westerly wind response to the heating is to be expected dynamically (see Section 6.8). Of all wind forcing, the ocean is most sensitive to that by zonal equatorial winds (see Sections 5.2 and 5.4). Thus Fig. 2.2(c) shows that the west/central equatorial Pacific region between about 150°E and 150°W is where *both* the ENSO forcing of the atmosphere *and* the ENSO forcing of the ocean are strongest.

Why should the interannual heating of the atmosphere be strongest in this west/central equatorial Pacific region? One possible line of reasoning is that a higher SST anomaly will cause anomalous heating of the overlying air which would then expand, become less dense and hence more buoyant than the surrounding air and consequently rise. Rising air cools and the water vapor in the air condenses and heats the air as latent heat is released. This heating causes more expansion of the air, further upward air movement, condensation of water vapor and further atmospheric heating. But if this line of reasoning were correct, we would expect the lowest OLR over the region of maximum SST anomaly. A comparison of Fig. 2.2(b) and (c) shows that this is not the case – as noted above, anomalous cloudiness (as measured by anomalous OLR) occurs in the west/central equatorial Pacific while the maximum anomalous SST occurs in the eastern equatorial Pacific almost a half hemisphere away. Clearly, anomalous cloud and atmospheric heating depend on more than just anomalous SST.

Observations of tropical OLR and SST (see Fig. 2.3) show that lower OLR (corresponding to increased cloudiness and rainfall) is much more likely for SST above about 27.5°C (see also Gadgil et al. 1984; Graham and Barnett 1987 and Sections 6.3–6.5). Thus, for SST anomalies to be effective in producing deep convection, the background SST must be at least 27.5°C. This result and Fig. 2.4 explain the ENSO atmospheric

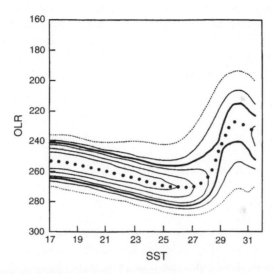

Fig. 2.3 Probability distribution of monthly mean OLR with respect to SST for 20°S–20°N. Contour values are successively 2% (dashed curve), 5, 7.5% (heavy solid curve), 10, 20 and 40%. The dotted curve denotes the OLR value with the highest probability. The probability of low OLR (and corresponding deep convection) increases dramatically above about 27.5°C. (From Zhang 1993.)

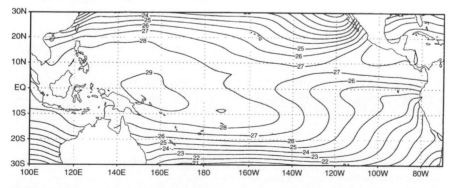

Fig. 2.4 Mean SST in the tropical Pacific in degrees centigrade. This figure was kindly provided by D. Stokes and R. Reynolds, National Meteorological Center.

heating anomaly pattern in Fig. 2.2(c). Although the SST anomalies are smaller in the west/central Pacific, they occur on top of SST greater than 27.5°C and produce anomalous convection while the larger SST anomalies in the eastern equatorial Pacific occur on top of water so cold that no deep atmospheric convective heating occurs.

What causes the equatorial SST anomalies in the west/central Pacific region 150°E–150°W where ENSO coupling between atmosphere and ocean is strongest? Analyses of observations and numerical models (Gill 1983; Fu et al. 1986; McPhaden and Picaut 1990; Picaut and Delcroix 1995; Picaut et al. 1996) suggest that zonal advection of water by ocean currents contributes substantially to the SST anomalies there. Specifically, suppose that a particle of water at the sea surface near the equator does not gain or lose heat to its surroundings but is moved eastward by an anomalous eastward ocean current. Since in the mean the SST steadily decreases toward the east in this region (Fig. 2.4), a positive SST anomaly and anomalous deep atmospheric convection result. We can think of this process as the anomalous current moving the warm water and convection eastward. An extreme example of this occurred in 1982–1983 when warm water from the western Pacific, westerly winds and deep atmospheric convection marched right across the equatorial Pacific. More typically, the warm water near the edge of the western Pacific warm pool only moves eastward as far as about 150°W before moving westward to about 160°E (Fig. 2.5). Like the extreme 1982–1983 episode, westerly winds and deep atmospheric convection follow the movement of the 28.5°C isotherm (see Fig. 6.1 of Chapter 6 and Fu et al. 1986).

Figure 2.5 shows that the zonal equatorial displacement of the eastern edge of the western Pacific warm pool is highly correlated with the Southern Oscillation Index (SOI). This index is defined here as the anomalous monthly surface atmospheric pressure at Tahiti divided by its standard deviation minus the anomalous monthly surface atmospheric pressure at Darwin divided by its standard deviation. The monthly pressure anomaly is the monthly pressure for a given calendar month minus the average value for that month; e.g., the anomalous pressure at Darwin in April 1997 would be the pressure at Darwin in April 1997 minus the average Darwin April pressure. In Fig. 2.5, why should the SOI be correlated with the zonal warm pool movement? Along the equator, winds blow from a high to a low pressure in order to transfer anomalous air mass and equalize pressure. During a warm ENSO episode, surface equatorial winds are westerly from about 150°E to 150°W (Fig. 2.2). Therefore, as shown in Fig. 2.2(a), surface pressure is high in the western equatorial Pacific (west of about 150°E) and low in the eastern equatorial Pacific (east of about 150°W). But the Tahiti surface pressure is indicative of equatorial surface pressure

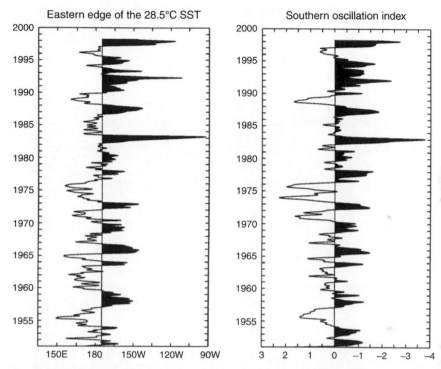

Fig. 2.5 Zonal displacement of the (left) 28.5°C SST averaged over 4°N–4°S and (right) Southern Oscillation Index (SOI) from 1950 to 1998. The SOI axis is inverted. The monthly 28.5°C zonal displacement has been smoothed with a 3-point running mean filter and the SOI has been filtered three times with this filter. (From Clarke et al. 2000.)

east of 150°W, and the Darwin atmospheric pressure is indicative of the equatorial pressure west of 150°E (Fig. 2.2(a)). Therefore, when the warm pool moves eastward, deep atmospheric convection moves eastward and anomalous equatorial westerly winds are induced along with anomalous high pressure in Darwin and anomalous low pressure in Tahiti, i.e., the SOI, being Tahiti minus Darwin anomalous surface pressure, is negative. This is in agreement with Fig. 2.5. Similar arguments can be given when the SOI is positive.

2.2.2. ENSO effects in the eastern Pacific

Figure 2.2(c) shows that north of the equator in the eastern Pacific, surface winds are roughly north–south and converge on approximately latitudes 4–5°N. Such a convergence of wind anomalies will produce

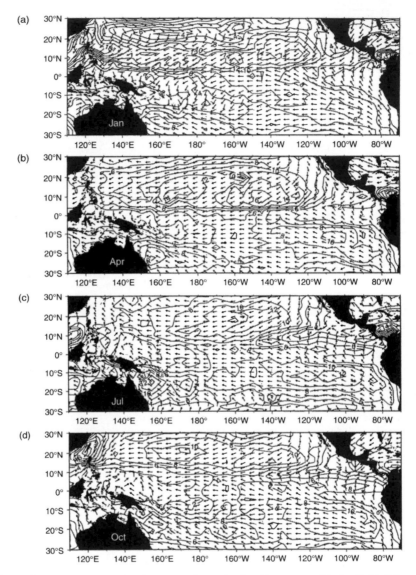

Fig. 2.6 'Pseudostress' for the tropical Pacific for (a) January (b) April (c) July and (d) October. Pseudostress is defined as the monthly average of $|u|u$ where, in the present context, u is the velocity of the wind 10 m above the water surface. The arrows, all of equal length, denote the direction of the pseudostress and contours denote the magnitude in $m^2 s^{-2}$. Note the strong convergence at about 5°N in the eastern Pacific in January, at about 5°N in April, at about 10°N in July and at about 10°N in October. This convergence is associated with the Intertropical Convergence Zone (ITCZ). This figure used data from the Comprehensive Ocean Atmosphere Data Set (COADS). (From Slutz et al. 1985.)

upward moving air from the surface at that latitude and hence clouds and rainfall. This is consistent with the approximately zonal band of decreased OLR at these latitudes (Fig. 2.2(c)).

Why does this anomalous band of surface wind convergence occur? Figure 2.6 shows that the southeast and northeast trade winds in the eastern tropical Pacific converge on a latitude band which is known as the Intertropical Convergence Zone (ITCZ). The ITCZ moves north and south seasonally and also interannually. When a warm ENSO episode occurs, the ITCZ moves southward (Ramage 1975; Hastenrath 1976; Rasmusson and Carpenter 1982: Deser and Wallace 1990; Clarke 1992), making the surface winds anomalously convergent near the equator and less convergent (anomalously divergent) at about 10°N (Fig. 2.2(d)).

Notice that the convergence near the ITCZ is at least as large as that along the equator between 160°E and 160°W, but the OLR anomaly and associated rainfall are much larger in the western equatorial Pacific. This is because the layer of air that converges over the western equatorial Pacific is deeper (Deser and Wallace 1990) and warmer; it therefore has more water vapor that releases more latent heat and gives rise to higher clouds on average.

Another salient feature of ENSO in the eastern equatorial Pacific is its strong anomalous SST (Fig. 2.2(b)). Such a strong SST signal is expected because of the temperature structure of the upper ocean in this region (Fig. 2.7). During El Niño, the thermocline, or region of strong vertical temperature gradient, is displaced downward, typically by about 20 m (Kessler 1990). Since the sea surface displacement is negligible compared to this, Fig. 2.7 suggests that this downward displacement will lead to an

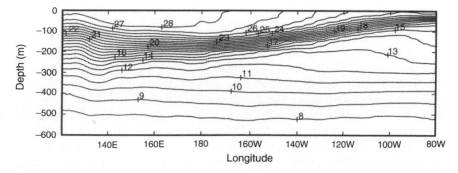

Fig. 2.7 Temperature (in °C) for a longitude–depth section along the equator in the Pacific. The temperature data are from the World Ocean Atlas of the United States National Oceanic and Atmospheric Administration.

invasion of warm water from the west which, because of the strong zonal SST gradient, will correspond to a sizable SST anomaly.

Rasmusson and Carpenter (1982) discovered that this SST anomaly, as well as surface wind, rainfall and surface pressure anomalies, is phase-locked to the calendar year. For example, maximum SST anomalies in the eastern equatorial Pacific usually occur around November. This is puzzling. Why should anomalies, which are calculated by removing the average seasonal cycle, have any seasonality? We will discuss phase-locking further in Section 2.3 and in detail in Chapter 8.

2.2.3. ENSO effects in the far western Pacific (120°E–140°E)

Notice that during a warm ENSO episode when the warm water and deep atmospheric convection move eastward along the equator, the western Pacific SST is lower than during a cold ENSO episode (compare Fig. 2.8(a) with (b) and Fig. 2.8(c) with (d)). Figures 2.2(c) and 2.9(c), which show a positive OLR anomaly over the western Pacific, indicate that the western Pacific is also drier. Fewer clouds and less rainfall near the equator also should result in anomalously easterly winds (see the dynamics discussed in Section 6.8) and in Fig. 2.9(c) these can be seen at the western part of the 'dry' (anomalously high OLR) region in the far western equatorial Pacific and eastern equatorial Indian Ocean. The easterlies are weaker and less extensive in Fig. 2.2(c), perhaps because of the influence of westerly winds remotely generated by the strong air–sea interaction between about 150°E and 150°W. In December–February, the intense interannual convection is slightly further to the east and south of the equator (see Fig. 2.9) and the easterly far western Pacific winds are more pronounced.

2.2.4. ENSO effects in the upper atmosphere

Figure 2.10 shows that the winds typically blow from the east in the lower atmosphere over most of the equatorial Pacific from near the coast of South America to the International Date Line and are in the opposite direction in the upper atmosphere. The air descends in the eastern equatorial Pacific and ascends in the west. As noted in Chapter 1, the equatorial circulation of surface easterlies, rising air in the western Pacific, upper level westerlies and descending air in the eastern Pacific was named the 'Walker Circulation' after Sir Gilbert Walker (Bjerknes 1969). When a warm ENSO episode occurs, as noted earlier, the warm pool moves eastward and so does the western equatorial Pacific rising air, thus shrinking the Walker Circulation. Typically, the far western Pacific (especially 120°E–140°E) is drier than usual, so the deep atmospheric convection is weaker and the upward velocity is smaller.

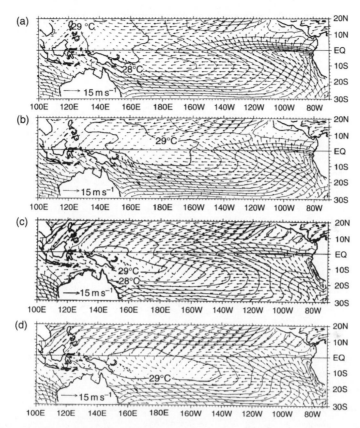

Fig. 2.8   (a) July–November total SST (contour interval = 1°C) and surface wind composites for a typical cold episode in the eastern equatorial Pacific. Composites were calculated by multiplying the regression coefficients in Fig. 2.2(b) by −0.8°C and adding the result to the average SST and wind fields for July–November. The temperature −0.8°C was chosen because it was the average SST anomaly for the Fig. 2.1 (a) index for the five coldest years. (b) July–November total SST and surface wind composites for a typical warm ENSO episode. In this case, the composites were obtained by multiplying the regression coefficients in Fig. 2.2(b) by +1.4°C and adding to the average SST and wind fields for July–November. By analogy with (a), +1.4°C is the average temperature anomaly for the Fig. 2.1(a) index for the five warmest years. (c) December–February total SST and surface wind composites for a typical cold ENSO episode. The composites were constructed as explained in (a) but by multiplying the December–February anomaly SST and wind fields in Fig. 2.9 by −0.8°C and adding these to the December–February average SST and wind fields. The temperature −0.8°C is the average anomaly for the five coldest years in Fig. 2.1(b). (d) December–February total SST and surface wind composites for a typical warm ENSO episode. In this case, the anomaly fields of Fig. 2.9 were multiplied by +1.5°C before adding to the mean December–February fields. The temperature +1.5°C is the average SST anomaly for the Fig. 2.1(b) index for the five warmest years. (From Deser and Wallace 1990.)

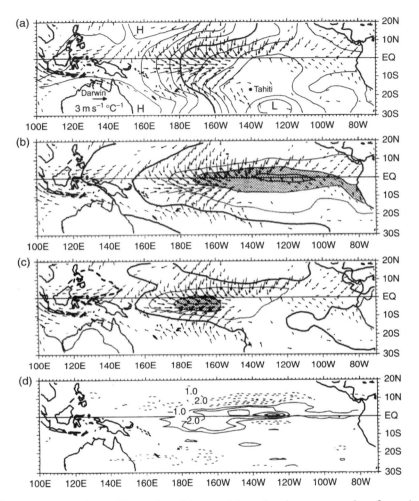

Fig. 2.9 Regression of December–February (a) sea level pressure and surface wind, (b) SST and surface wind, (c) OLR and surface wind and (d) surface wind divergence upon the equatorial SST index in Fig. 2.1(b). Plotting convention as in Fig. 2.2. (Redrawn from Deser and Wallace 1990.)

Warm and cold ENSO episodes affect the equatorial Pacific atmospheric temperature as well as the winds. In fact, on average, the tropical troposphere (approximately the lower 17 km of the atmosphere) heats and cools by about 1/2°C all around the earth during warm and cold ENSO episodes (Horel and Wallace 1981; Yulaeva and Wallace 1994). We will discuss this further in Chapter 9.

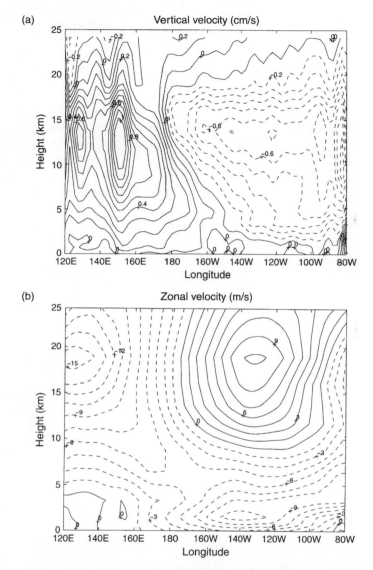

Fig. 2.10   (a) Mean atmospheric vertical velocity (cm s^{-1}) along the equator in a height by longitude vertical plane for the Pacific. Dashed contours denote negative vertical velocity (descending air). The plot is derived from data supplied by the NOAA–CIRES Climate Diagnostics Center; (b) as for (a) but for eastward velocity in m s^{-1}. The two plots show the general structure of the Pacific Walker Circulation: rising air in the west and sinking air in the east connected by the trade wind surface easterlies and westerly flow aloft.

2.3. TIME SERIES STRUCTURE OF THE ENSO VARIABLES

In Section 2.2, we examined the spatial patterns of various important physical quantities in ENSO and their relationship to each other. Next we will consider the time-varying structure of ENSO by examining two widely used ENSO indices – the Tahiti minus Darwin SOI discussed earlier in Section 2.2.1 and the El Niño index NINO3.4. The latter monthly index is defined as the SST anomaly averaged over the east-central equatorial Pacific region 5°S–5°N, 170°W–120°W.

Figure 2.5 shows that ENSO is irregular with much of the variability on the 'interannual' time scale, i.e., a typical periodicity of a few years. This is clear also from the spectrum in Fig. 2.11, which shows that most of the SOI variability occurs for periods between about 2 and 8 years.

A closer examination of the SOI and NINO3.4 shows that these time series have more structure than random interannual variability.

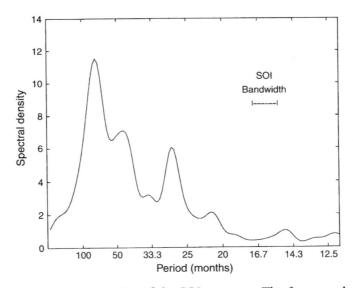

Fig. 2.11 Low-frequency portion of the SOI spectrum. The frequency bandwidth is plotted and there are 10 degrees of freedom. The abscissa is linear in frequency. The spectrum is based on approximately 121 years of monthly SOI data (January 1876–March 1997) generously supplied by Dr. R. Allan. (From Clarke and Van Gorder 1999.)

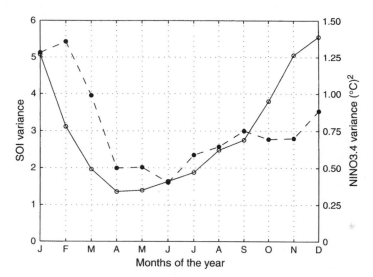

Fig. 2.12 Variance of the Tahiti–Darwin SOI (dashed curve, closed circles) and the El Niño index NINO3.4 (open circles, solid curve) as a function of calendar month.

Paradoxically, even though these indices are anomaly time series and therefore have no annual cycle, their variance and persistence are strongly phase-locked to the annual cycle! For example, Fig. 2.12 shows that NINO3.4 and the SOI have minimum variance in the Northern Hemisphere spring/early summer, while Fig. 2.13 shows that the SOI is much more highly correlated across the fall than across the spring, i.e., the SOI is much more persistent across the fall. This persistence property can be seen in more detail in Fig. 2.14(a), which shows that there is a precipitous drop in SOI persistence in the Northern Hemisphere Spring. The El Niño SST index NINO3.4 is even more persistent across the fall and has a bigger drop in persistence in the spring (Fig. 2.14(b)). The strong persistence of NINO3.4 across the fall and winter makes it easy to predict ENSO from the end of July to January the following year; the correlation of the July value of NINO3.4 with any of the next 6 months is greater than 0.8.

The large drop in persistence during the Northern Hemisphere Spring is known as the spring persistence barrier. First noticed by Walker (1924), it has also been documented by Walker and Bliss (1932, 1937), Troup (1965), Wright (1977), Webster and Yang (1992) and Webster (1995). We will discuss this and other calendar year phase-locked ENSO properties in Chapter 8.

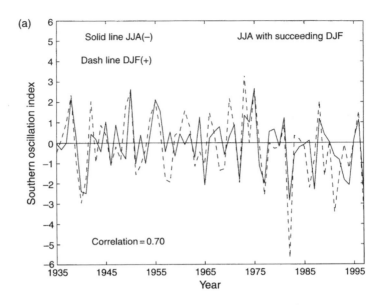

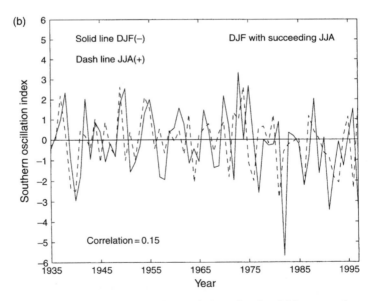

Fig. 2.13 (a) Time series and lagged correlations for the SOI averaged over June, July and August (JJA) and the SOI averaged over the following December, January and February (DJF). (b) The DJF values of the SOI correlated with the following JJA values. (Adapted from Webster and Yang 1992.)

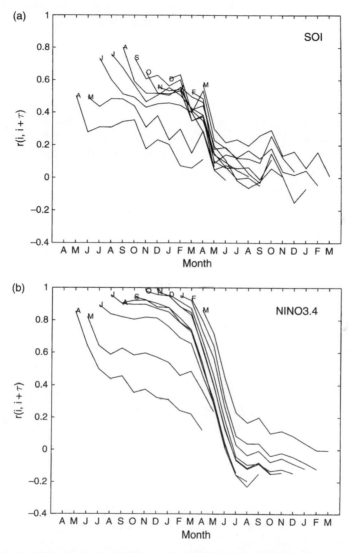

Fig. 2.14 (a) The correlation between the SOI for calendar month i and calendar month $i+\tau$ for $i = 1, 2, \ldots, 12$ and $\tau = 1, 2, \ldots, 12$. For example, for $i = 1$ and $i+\tau = 3$, the correlation is between the January SOI time series and the March SOI time series. In the figure, each curve is labeled by the first letter of its starting month i, and the curve begins with the correlation between the i and $i+1$ calendar month time series. For example, the first point on the curve labeled 'S' corresponds to the correlation between the September and October SOI time series (Redrawn from Clarke and Van Gorder 1999; see also Webster and Yang 1992). (b) As for (a) but for NINO3.4 (average SST anomaly for the region 5°N–5°S, 170°W to 120°W) instead of the SOI.

2.4. A Physical Explanation for El Niño

Equatorial oceans are strongly heated near the surface and stirred by the trade winds. Consequently, to a first approximation, the equatorial Pacific Ocean consists of a shallow, warm surface layer of water that varies in depth from a few tens of meters to 100 m or more. This layer overlies a region of sharply decreasing temperature (the thermocline) and then a much thicker layer in which the temperature decreases gradually to the bottom at about 4000 m (see Fig. 2.7). Notice that the warm surface layer shoals from the central to the eastern Pacific and that, as a result, the upper part of the thermocline is mixed into the surface layer in the eastern Pacific. This causes the SST to be increasingly lower as one moves eastward along the equator (Fig. 2.4). The shoaling thermocline is due to the equatorial trades; they mainly affect the warm, surface water above the thermocline and, since they blow toward the west, force the warm surface water over to the west and make the surface layer thicker there.

The shoaling thermocline has profound effects on marine life in the eastern equatorial Pacific. At the base of the food chain are microscopic marine plants (phytoplankton) which need light for photosynthesis and inorganic plant nutrients such as nitrate, phosphate and silicate. Light only penetrates about 50 m into the ocean, and the major inorganic nutrient reservoir in the ocean occurs beneath the thermocline. Therefore, in the eastern equatorial Pacific where, on average, the thermocline is near the surface, phytoplankton, zooplankton (small marine animals), birds, fish and marine mammals are plentiful. However, when El Niño occurs, the equatorial trade winds weaken and the upper warm surface layer adjusts to the weaker winds; a shallower thermocline in the western Pacific and a deeper thermocline in the eastern Pacific result. The adjustment takes place by equatorial Kelvin and Rossby ocean wave dynamics, a subject to be taken up in Chapter 5. In the eastern equatorial Pacific, a deeper thermocline results in warmer water (see Fig. 2.2(b)), fewer nutrients available for phytoplankton and consequently a marked decrease in marine plant (see Figs 2.15–2.17) and animal life.

Thus remote variations in equatorial trade wind strength (see Fig. 2.2(b)), roughly a quarter of the way around the earth distant from the eastern equatorial Pacific, profoundly affect the physical and chemical ocean environment there and dramatically affect the marine life. The idea that El Niño is remotely forced by a decrease in equatorial trade wind strength was first proposed by Wyrtki (1975) and Godfrey (1975).

But what causes the wind variations in the first place? According to the discussion in Section 2.2, these are driven by anomalous deep atmospheric convective heating, which is in turn driven by SST anomalies which are in

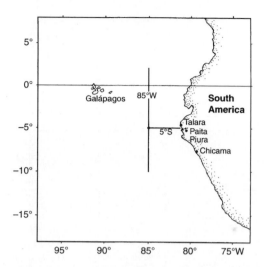

Fig. 2.15 Equatorial Pacific Ocean showing location of the cross–equatorial profile of Fig. 2.16 and the cross-shelf profile of 2.17. (Redrawn from Barber and Chavez 1983.)

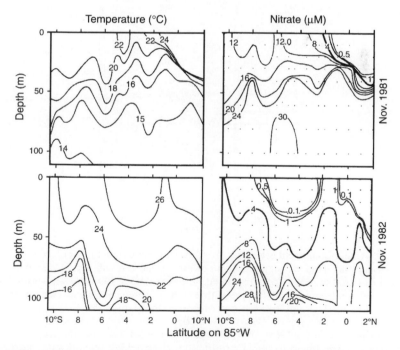

Fig. 2.16 Cross-equatorial profiles of temperature and nutrient nitrate along a transect at 85° W from 2°N to 10°S (see Fig. 2.15). In November 1981 (top two panels), conditions were normal; by November 1982 (bottom two panels), there was an onset of warmer, low-nutrient, anomalous conditions. (Redrawn from Barber and Chavez 1983.)

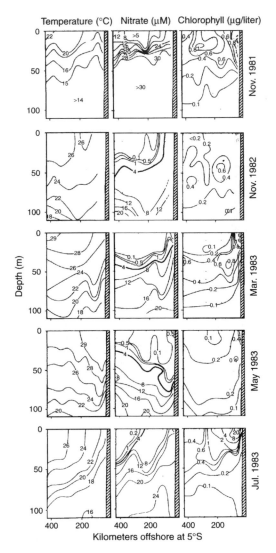

Fig. 2.17 Cross-shelf profiles along a 5°S transect from 85°W to Paita (see Fig. 2.15). November 1981 (top three panels) conditions were normal; by November 1982 (second row of three panels), there was an onset of warmer, low–nutrient and reduced chlorophyll conditions; by March 1983 (third row of three panels), surface waters had become even warmer and lower in nutrients and chlorophyll; the surface waters continued to warm to May 1983 (fourth row of three panels) and surface water nutrients and chlorophyll remained low; by July 1983 (last row of three panels), cold nutrient-rich waters had returned to the nearshore coastal upwelling region, and chlorophyll concentration increased there. Offshore, the water was still warm and low in nutrients and chlorophyll. (Redrawn from Barber and Chavez 1983.)

turn driven by the wind. The argument is circular because ENSO is a fully coupled ocean–atmosphere problem. Coupled ocean–atmosphere dynamics, including two popular paradigms, will be discussed in Chapter 7. But before we can discuss coupled dynamics, we must discuss ocean dynamics (Chapters 3–5) and then the atmospheric response to anomalous SST heating (Chapter 6).

 REFERENCES

Barber, R. T., and F. P. Chavez, 1983: Biological consequences of El Niño. *Science*, **222**(4629), 1203–1210.

Bjerknes, J., 1969: Atmospheric teleconnections from the equatorial Pacific. *Mon. Weather Rev.*, **97**, 163–172.

Clarke, A. J., 1992: Low frequency reflections from a nonmeridional eastern ocean boundary and the use of coastal sea level to monitor eastern Pacific equatorial Kelvin waves. *J. Phys. Oceanogr.*, **22**, 163–183.

Clarke, A. J., and S. Van Gorder, 1999: On the connection between the boreal spring southern oscillation persistence barrier and the tropospheric biennial oscillation. *J. Climate*, **12**(2) 610–620.

Clarke, A. J., J. Wang, and S. Van Gorder, 2000: A simple warm-pool displacement ENSO model. *J. Phys. Oceanogr.*, **30**(7), 1679–1691; Corrigendum *JPO*, **30**(12), 3271.

Deser, C., and J. M. Wallace, 1990: Large-scale atmospheric circulation features of warm and cold episodes in the tropical Pacific. *J. Climate*, **3**, 1254–1281.

Fu, C., H. F. Diaz, and J. O. Fletcher, 1986: Characteristics of the response of sea surface temperature in the central Pacific associated with warm episodes of the Southern Oscillation. *Mon. Weather Rev.*, **114**, 1716–1738.

Gadgil, S., P. V. Joseph, and N. V. Joshi, 1984: Ocean–atmosphere coupling over monsoon regions. *Nature* (London), **312**, 141–143.

Gill, A. E., 1983: An estimation of sea level and surface-current anomalies during the 1972 El Niño and consequent thermal effects. *J. Phys. Oceanogr.*, **13**, 586–606.

Godfrey, J. S., 1975: On ocean spin-down. I. A linear experiment. *J. Phys. Oceanogr.*, **5**, 399–409.

Graham, N. E., and T. P. Barnett, 1987: Sea surface temperature, surface wind divergence and convection over tropical oceans. *Science*, **238**, 657–659.

Hastenrath, S., 1976: Variations in low-latitude circulation and extreme climatic events in the tropical Americas. *J. Atmos. Sci.*, **33**, 202–215.

Horel, J. D., and J. M. Wallace, 1981: Planetary-scale atmospheric phenomena associated with the Southern Oscillation. *Mon. Weather Rev.*, **109**, 813–829.

Kessler, W. S., 1990: Observations of long Rossby waves in the northern tropical Pacific. *J. Geophys. Res.*, **95**, 5183–5217.

McPhaden, M. J., and J. Picaut, 1990: El Niño-Southern Oscillation displacement of the western equatorial Pacific warm pool. *Science*, **250**, 1385–1388.

Picaut, J., and T. Delcroix, 1995: Equatorial wave sequence associated with warm pool displacements during the 1986–1989 El Niño-La Niña. *J. Geophys. Res.*, **100**, 18393–18408.

Picaut, J., M. Ioualalen, C. Menkes, T. Delcroix, and M. J. McPhaden, 1996: Mechanism of the zonal displacements of the Pacific warm pool: Implications for ENSO. *Science*, **274**, 1486–1489.

Ramage, C. S., 1975: Preliminary discussion of the meteorology of the 1972–73 El Niño. *Bull. Am. Meteorol. Soc.*, **56**, 234–242.

Rasmusson, E. M., and T. H. Carpenter, 1982: Variations in sea surface temperature and surface wind fields associated with the Southern Oscillation/El Niño. *Mon. Weather Rev.*, **110**, 354–384.

Slutz, R. J., S. D. Woodruff, R. L. Jenne, and P. M. Steurer, 1985: A comprehensive ocean-atmosphere data set. *Bull. Am. Meteorol Soc.*, **68**, 1239–1250.

Troup, A. J., 1965: The Southern Oscillation. *Q. J. R. Meteorol. Soc.*, **91**, 490–506.

Walker, G. T., 1924: Correlation in seasonal variations in weather, IX. A further study of world-weather. *Mem. Indian Meteorol. Dep.* **24**, 275–332.

Walker, G. T., and E. W. Bliss, 1932: World weather V. *Mem. R. Meteorol. Soc.*, **4**, 53–84.

Walker, G. T., and E. W. Bliss, 1937: World weather VI. *Mem. R. Meteorol. Soc.*, **4**, 119–139.

Webster, P. J., 1995: The annual cycle and the predictability of the tropical coupled ocean–atmosphere system. *Meteorol. Atmos. Phys.*, **56**, 33–55.

Webster, P. J., and S. Yang, 1992: Monsoon and ENSO: Selectively interactive systems. *Q. J. R. Meteorol. Soc.*, **118**, 877–926.

Wright, P. B., 1977: The Southern Oscillation – Patterns and mechanisms of the teleconnections and the persistence. *Hawaii Institute of Geophysics Report*, HIG-77-13.

Wyrtki, K., 1975: El Niño – The dynamic response of the equatorial Pacific Ocean to atmospheric forcing. *J. Phys. Oceanogr.*, **5**, 572–584.

Yulaeva, E., and J. M. Wallace, 1994: The signature of ENSO in global temperature and precipitation fields derived from the microwave sounding unit. *J. Climate*, **7**, 1719–1736.

Zhang, C., 1993: Large-scale variability of atmospheric deep convection in relation to sea surface temperature in the tropics. *J. Climate*, **6**, 1898–1913.

Equatorial Ocean Waves

Contents

3.1. Overview

As noted at the conclusion of the previous chapter, ENSO is a coupled ocean–atmosphere process. Equatorial winds force the equatorial ocean and cause the SST to change. This change in SST then changes the heating of the atmosphere which then changes the wind which then changes the SST, etc. Before we can understand the coupled dynamics we must understand how the ocean responds to atmospheric forcing and how the atmosphere responds to changes in SST. In the next three chapters we will focus on the ocean dynamics.

I assume that readers of this chapter have a basic knowledge of the equations of fluid mechanics on a rotating earth. Only linear theory will be considered since the wind-driven ocean flows relevant to ENSO are approximately linear. In Section 3.2, we will analyze a linear ocean model that is of constant depth, has horizontal stratification and is forced by the wind. For such an ocean it is useful to think in terms of 'vertical modes' and these are introduced in Section 3.3. Basic to the ocean response to the wind are equatorial ocean waves as they can carry energy over enormous distances and are directly responsible for El Niño. Section 3.4 treats equatorial wave theory. The reflection of equatorial waves is important both to ENSO dynamics and the remote influence of ENSO variability on mid-latitude coastal ocean physics and biology (see Sections 12.4–12.6). Equatorial wave reflection from the western and eastern Pacific Ocean boundaries is covered in Chapter 4. Chapter 5 shows how the wind-forced equatorial ocean low-frequency variability can be interpreted as a sum of wind-forced equatorial waves. The last part of Chapter 5 uses the concepts developed in Chapters 3–5 to explain how El Niño is generated remotely at the South American coast by equatorial winds about 10 000 km away.

3.2. The Linear Stratified Ocean Model

Consider an ocean model with a right-handed system of axes with the x-axis eastward, the y-axis northward and the z-axis vertically upward. The equation of the equator is $y=0$ and the sea surface at rest is $z=0$. For a state of rest the density and pressure are functions of z alone and are in hydrostatic balance:

$$\frac{\mathrm{d}p_0}{\mathrm{d}z} = -g\rho_0(z) \tag{3.1}$$

where g is the acceleration due to gravity and $p_0(z)$ and $\rho_0(z)$ refer, respectively, to pressure and density in a state of rest. Under small perturbations to this state of rest, the linearized momentum equations are

$$u_t - fv = -p'_x/\rho_0 + X_z/\rho_0 \tag{3.2}$$

$$v_t + fu = -p'_y/\rho_0 + Y_z/\rho_0 \tag{3.3}$$

$$0 = -p'_z - g\rho' \tag{3.4}$$

and the incompressibility condition and equation of mass reduce to

$$u_x + v_y + w_z = 0 \tag{3.5}$$

$$\rho'_t + w\rho_{0z} = 0. \tag{3.6}$$

In (3.2)–(3.6) u, v and w refer to velocity components in the eastward (x), northward (y) and upward (z) directions, f is the Coriolis parameter and p' and ρ' are pressure and density perturbations due to the motion. In

(3.4), I have neglected the local acceleration of the vertical velocity and the vertical Coriolis acceleration. This is reasonable for the motion relevant to our case, viz., flow in which the horizontal scale is much greater than the water depth or other shorter vertical scales related to the depth-varying stratification.

In (3.2) and (3.3) the vector (X, Y) is the tangential Reynolds stress acting between horizontal planes and takes a prescribed value, (τ^x, τ^y), the wind stress at the surface. Readers unfamiliar with Reynolds stress should consult, e.g., Kundu (1990). Note that the complete force/unit mass vector due to the Reynolds stress is actually

$$-\left(\frac{\partial}{\partial x}(\overline{u''u''}) + \frac{\partial}{\partial y}(\overline{u''v''}) + \frac{\partial}{\partial z}(\overline{u''w''}), \quad \frac{\partial}{\partial x}(\overline{v''u''}) + \frac{\partial}{\partial y}(\overline{v''v''}) + \frac{\partial}{\partial z}(\overline{v''w''}) \right)$$

where u'', v'' and w'' are turbulent velocity fluctuations with zero means. However, because of the stable horizontal stratification, the z-scale associated with the Reynolds stress is much smaller than the horizontal scales. The comparatively small z-scale and the assumption that $(\overline{u''u''})$, $(\overline{u''v''})$... and $(\overline{u''w''})$ are of the same order together imply that one can approximate the complete force/unit mass vector by

$$-\frac{\partial}{\partial z}(\overline{u''w''}), \quad -\frac{\partial}{\partial z}(\overline{v''w''}) = (X_z, Y_z)$$

(see (3.2) and (3.3)).

Solution of (3.2)–(3.6) requires a knowledge of how (X, Y) varies with depth. Because of the stable horizontal stratification the stress is negligible except in a thin well-mixed layer near the surface. It will be assumed that this well-mixed layer, of depth H_{mix}, moves as a slab, i.e., that the horizontal velocity (u, v) is independent of z over the mixed layer. Since the horizontal currents in the mixed layer are balanced by the force vector $(X_z/\rho_0, Y_z/\rho_0)$, the assumption of depth-independent, wind-driven currents in the mixed layer is equivalent to assuming that X and Y are linear functions of z in the mixed layer, falling from their surface values to zero at the bottom of the mixed layer. For this model, therefore, the vertical derivative (X_z, Y_z) of the stress is given by

$$(X_z, Y_z) = \begin{cases} (\tau^x, \tau^y)/H_{\text{mix}} & -H_{\text{mix}} \leq z \leq 0 \\ 0 & z < -H_{\text{mix}} \end{cases}. \tag{3.7}$$

By eliminating ρ' between (3.4) and (3.6) we deduce

$$w = -p'_{zt}/\rho_0 N^2 \tag{3.8}$$

where N, the 'buoyancy frequency', is related to the buoyancy of the water and is defined from

$$N^2 = -g\rho_0^{-1}\, d\rho_0/dz. \tag{3.9}$$

Substitution of (3.8) into (3.5) gives

$$u_x + v_y + \left(-p'_{zt}/\rho_0 N^2\right)_z = 0 \tag{3.10}$$

so our ocean model is governed by the three equations (3.2), (3.3) and (3.10) for the dependent variables u, v and p'.

Note that the third term on the left-hand side of (3.10) is

$$\left(-p'_{zt}/\rho_0 N^2\right)_z = \rho_0^{-1}\left(-p'_{zt}/N^2\right)_z + \rho_0^{-2}\,\rho_{0z}\,p'_{zt}/N^2. \tag{3.11}$$

In the ocean, density typically changes by less than $1/2\%$ from surface to bottom so the z-scale ρ_0/ρ_{0z} is enormous compared to the ocean depth. But N^2 and p' typically vary on scales of the order of the ocean depth or smaller so in (3.11) the second term on the right-hand side is negligible compared to the first. Thus with small error (3.11) may be written as

$$\left(-p'_{zt}/N^2\rho_0\right)_z = \rho_*^{-1}\left(-p'_{zt}/N^2\right)_z \tag{3.12}$$

where we have also used $\rho_0^{-1} = \rho_*^{-1}$ with ρ_* being the mean water density. Equation (3.12) is a version of the 'Boussinesq approximation.' Using (3.12), the third equation (3.10) governing the flow reduces to

$$u_x + v_y + \rho_*^{-1}\left(-p'_{zt}/N^2\right)_z = 0. \tag{3.13}$$

The governing equations (3.2), (3.3) and (3.13) are subject to free surface and bottom boundary conditions. The linearized free surface condition is

$$w = \eta'_t \qquad \text{on} \quad z = 0 \tag{3.14a}$$

where $\eta'(x, y, t)$ is the height of the sea level above $z = 0$. The ocean is assumed to be of constant depth H and so the bottom boundary condition is

$$w = 0 \qquad \text{on} \quad z = -H. \tag{3.14b}$$

We wish to write the above boundary conditions in terms of p' since w and η' do not appear explicitly in the governing equations (3.2), (3.3) and (3.13). Integrating the equation for hydrostatic balance from some point z close to the surface $z = \eta'$ gives

$$\int_z^{\eta'} p_z\, dz = \int_z^{\eta'} -g\rho\, dz \approx \int_z^{\eta'} -g\rho_*\, dz$$

since we take z close enough to the surface that ρ departs negligibly from ρ_*. Since the pressure at the ocean surface matches the atmospheric pressure p_a we have that the total pressure p at depth $|z|$ is

$$p = p_a + g\rho_* \left(\eta' - z \right), \qquad (3.15)$$

i.e., the ocean pressure at depth $|z|$ is equal to a sum of the atmospheric pressure and the pressure due to the weight of water above that depth. Because p is a sum of $p_0(z)$ and p', differentiating (3.15) with respect to t and substituting for η' in (3.14a) gives

$$w = \frac{p'_t - p_{at}}{g\rho_*} = \frac{p'_t}{\rho_* g}$$

where we have ignored atmospheric forcing effects since they are negligible. Using the result $w = -p'_{zt}/\rho_* N^2$ (see (3.8) with $\rho_0 \approx \rho_*$) we can then express the free surface condition in terms of p' as

$$p'_{zt}/N^2 + p'_t/g = 0 \qquad \text{on} \quad z = 0 \qquad (3.16)$$

and the bottom boundary condition as

$$p'_{zt} = 0 \qquad \text{on} \quad z = -H. \qquad (3.17)$$

3.3. The Description of the Constant Depth Ocean Dynamics in Terms of Vertical Modes

3.3.1. The continuously stratified case

The ocean model described in Section 3.2 involves four independent variables x, y, z and t and two z boundary conditions at the ocean surface and bottom. It is helpful to separate the model into 'vertical modes' which are to be described below. Each mode automatically satisfies the free surface and bottom boundary conditions. The z-dependent part of the solution is known and so the problem reduces to one involving three instead of four independent variables.

To separate into vertical modes write

$$u = \sum_{n=0}^{\infty} u_n(x, y, t)F_n(z), \quad v = \sum_{n=0}^{\infty} v_n(x, y, t)F_n(z), \quad \frac{p'}{\rho_*} = \sum_{n=0}^{\infty} p_n(x, y, t)F_n(z),$$

$$\frac{X_z}{\rho_*} = \sum_{n=0}^{\infty} X_n(x, y, t)F_n(z), \quad \frac{Y_z}{\rho_*} = \sum_{n=0}^{\infty} Y_n(x, y, t)F_n(z) \qquad (3.18)$$

where each vertical mode n is proportional to the eigenfunction $F_n(z)$ which satisfies the Sturm–Liouville problem:

$$\left(\frac{F_z}{N^2}\right)_z + \frac{F}{c^2} = 0 \qquad (3.19)$$

$$\frac{F_z}{N^2} + \frac{F}{g} = 0 \qquad \text{on} \quad z = 0 \qquad (3.20)$$

$$F_z = 0 \qquad \text{on} \quad z = -H. \qquad (3.21)$$

The eigenfunctions F_n form a complete set and are orthogonal in the sense that

$$\int_{-H}^{0} F_m F_n \, dz = 0 \qquad (m \neq n). \qquad (3.22)$$

The eigenfunctions are determined up to an arbitrary multiplicative constant and we will choose to normalize them by writing

$$F_n(0) = 1. \qquad (3.23)$$

It is also convenient to order the eigenfunctions so that the eigenvalues c_n decrease monotonically with n. In that case $c_n \rightarrow$ constant $(1/n)$ as $n \rightarrow \infty$ (see, e.g., Appendix B of Clarke and Van Gorder 1986). Because the density difference across the sea surface is so much greater than the density difference within the ocean, one mode has quite a different character from the others. This is the barotropic mode ($n = 0$) for which, to a good approximation

$$F_0(z) = 1 - \int_{-H}^{z} \frac{N^2}{c_0^2}(z_* + H)dz_* \approx 1 \text{ and } c_0 = (gH)^{1/2}. \qquad (3.24)$$

The above result can be checked by substitution into (3.19)–(3.21).

The vertical mode expressions (3.18) must satisfy the governing equations (3.2), (3.3) and (3.13) as well as the surface and bottom boundary conditions (3.16) and (3.17). The latter are automatically satisfied because of the eigenfunction properties (3.20) and (3.21). Substitution of (3.18) into the governing equations gives

$$\sum_{n=0}^{\infty} F_n(z)\left[u_{nt} - fv_n\right] = \sum_{n=0}^{\infty}(-p_{nx} + X_n)F_n(z)$$

$$\sum_{n=0}^{\infty} F_n(z)\left[v_{nt} + fu_n\right] = \sum_{n=0}^{\infty}(-p_{nx} + Y_n)F_n(z)$$

$$\sum_{n=0}^{\infty}\left(\frac{-F_{nz}}{N^2}\right)_z \left[p_{nt}\right] + F_n(u_{nx} + v_{ny}) = 0.$$

Using $(-F_{nz}/N^2)_z = F_n/c_n^2$ from (3.19), then multiplying the above 3 equations by $F_m(z)$ and integrating over the depth implies, from (3.22), that for each mode n

$$u_{nt} - fv_n = -p_{nx} + X_n \tag{3.25}$$

$$v_{nt} + fu_n = -p_{ny} + Y_n \tag{3.26}$$

$$\frac{p_{nt}}{c_n^2} + u_{nx} + v_{ny} = 0 \tag{3.27}$$

where, by (3.7) and (3.18)

$$X_n = \frac{b_n \tau^x}{\rho_*}, \quad Y_n = \frac{b_n \tau^y}{\rho_*}, \tag{3.28a}$$

$$b_n = \int_{-H_{mix}}^{0} F_n \, dz \bigg/ \left(H_{mix} \int_{-H}^{0} F_n^2 \, dz\right). \tag{3.28b}$$

3.3.2. The two-layer stratification case

Sometimes it is convenient to consider an ocean model with two layers of fluid of constant density (see Fig. 3.1). The linear equations for such a model can be found by integrating with respect to z the momentum equations (3.2) and (3.3) and the continuity condition (3.5). One finds, since the horizontal velocity is assumed independent of z, that

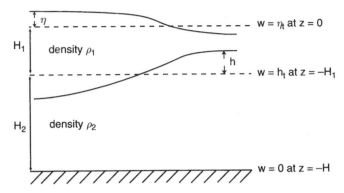

Fig. 3.1 The two-layer ocean model consisting of an upper layer of density ρ_1 and a lower layer of density ρ_2 with $(\rho_2 - \rho_1)/\rho_2 \ll 1$. At rest the upper layer has a depth H_1 and the lower layer a depth H_2. The total water depth $H = H_1 + H_2$. The sea level η is the upward displacement of the sea surface from its position of rest and h is the upward displacement of the interface between the upper and lower layers. The model is not drawn to scale; usually $H_1 \ll H$ and $\eta \ll h$.

$$u_t^{(1)} - fv^{(1)} = \frac{-p_x^{(1)}}{\rho_1} + \frac{\tau^x}{\rho_1 H_1} \tag{3.29}$$

$$v_t^{(1)} + fu^{(1)} = \frac{-p_y^{(1)}}{\rho_1} + \frac{\tau^y}{\rho_1 H_1} \tag{3.30}$$

$$\eta_t' - h_t + H_1 \left(u_x^{(1)} + v_y^{(1)} \right) = 0 \tag{3.31}$$

$$u_t^{(2)} - fv^{(2)} = \frac{-p_x^{(2)}}{\rho_2} \tag{3.32}$$

$$v_t^{(2)} + fu^{(2)} = \frac{-p_y^{(2)}}{\rho_2} \tag{3.33}$$

$$h_t + H_2 \left(u_x^{(2)} + v_y^{(2)} \right) = 0 \tag{3.34}$$

where, in this case the (1) and (2) superscripts refer to the upper and lower layer velocities, respectively, and h is the upward displacement of the interface between the two layers.

The pressures for the upper and lower layers can be found in terms of η and h from the hydrostatic relation

$$p_z = -g\rho . \tag{3.35a}$$

Since $\rho = \rho_1$ in the upper layer one has, by integrating from some point z in the upper layer to the surface,

$$p_a - p^{(1)}(x, y, z, t) = -g\rho_1(\eta' - z) \tag{3.35b}$$

where we have used the boundary condition that $p^{(1)}$ must match the atmospheric pressure p_a at the surface $z = \eta'$. Applying the horizontal gradient operator ∇ to (3.35b) and rearranging gives

$$\nabla p^{(1)}/\rho_1 = \nabla(g\rho_1\eta' + p_a)/\rho_1 = g\nabla\eta \tag{3.36a}$$

where η, the 'adjusted' sea level is given by

$$\eta = \eta' + p_a'/\rho_1 g \tag{3.36b}$$

with p_a' representing the departure of surface atmospheric pressure from its value when the atmosphere is at rest. Typically the adjustment $p_a'/\rho_1 g$ is small compared with η', consistent with our neglect of p_a' earlier in this chapter.

A similar integration from z to the surface when z is in the lower layer results in

$$p_a - p^{(2)}(x, y, z, t) = -g\rho_1(\eta' + H_1 - h) - g\rho_2(-H_1 + h - z)$$

and hence, using (3.36b) that

$$\nabla p^{(2)}/\rho_2 = g\nabla h(\rho_2 - \rho_1)/\rho_2 + g\rho_1\nabla\eta/\rho_2 = g\nabla((1 - \varepsilon)\eta + \varepsilon h) \tag{3.36c}$$

where

$$\varepsilon = (\rho_2 - \rho_1)/\rho_2. \tag{3.36d}$$

Equations (3.36) thus close the set of two-layer equations (3.29)–(3.34) – six equations in six unknowns (η, h, $u^{(1)}$, $v^{(1)}$, $u^{(2)}$, $v^{(2)}$). As in the continuously stratified case, the atmospheric pressure forcing $\partial p_a'/\partial t$ is negligible and so η' can be replaced by η in (3.31).

These six equations are really a special (degenerate) case of (3.2)–(3.6) with $N^2(z)$ being proportional to a δ function at $z = -H_1$. This case is degenerate in that the infinite set of modes collapses to just two: the barotropic mode and a single baroclinic mode. The modal equations in the form (3.25)–(3.27) can be found as follows. Add α times the three lower layer equations (3.32)–(3.34) to the corresponding three upper layer equations (3.29)–(3.31) to get, using (3.36a,c)

$$u_t - fv = -g([1 + \alpha(1 - \varepsilon)]\eta + \alpha\varepsilon h)_x + \frac{\tau^x}{\rho_1 H_1} \tag{3.37a}$$

$$v_t + fu = -g([1 + \alpha(1 - \varepsilon)]\eta + \alpha\varepsilon h)_y + \frac{\tau^y}{\rho_1 H_1} \tag{3.37b}$$

$$\left[\frac{\eta}{H_1} + \left(\frac{\alpha}{H_2} - \frac{1}{H_1}\right) h\right]_t + u_x + v_y = 0 \tag{3.37c}$$

where $(u, v) = (u^{(1)} + \alpha u^{(2)}, v^{(1)} + \alpha v^{(2)})$. The above equations have the same form as the modal equations (3.25)–(3.27) provided that

$$\frac{p}{c^2} = \frac{\eta}{H_1} + \left(\frac{\alpha}{H_2} - \frac{1}{H_1}\right) h$$

and

$$p = g\left([1 + \alpha(1 - \varepsilon)] \eta + \alpha\varepsilon h\right).$$

Elimination of p implies

$$\frac{g}{c^2} [\{1 + \alpha(1 - \varepsilon)\} \eta + \alpha\varepsilon h] = \frac{\eta}{H_1} + \left(\frac{\alpha}{H_2} - \frac{1}{H_1}\right) h.$$

Since this is true for all η and h, the coefficients of η and h must be equal and so

$$\frac{g}{c^2} [1 + \alpha(1 - \varepsilon)] = \frac{1}{H_1}, \quad \frac{g}{c^2} [\alpha\varepsilon] = \frac{\alpha}{H_2} - \frac{1}{H_1}.$$

Elimination of g/c^2 gives

$$\left(\frac{\alpha}{H_2} - \frac{1}{H_1}\right) \left(\frac{1}{\alpha\varepsilon}\right) (1 + \alpha(1 - \varepsilon)) = \frac{1}{H_1},$$

i.e.,

$$\alpha^2 \left[(1 - \varepsilon)\frac{H_1}{H_2}\right] + \alpha \left[\frac{H_1}{H_2} - 1\right] - 1 = 0. \tag{3.38}$$

For small ε, using $(H_1/H_2 - 1)^2 + 4H_1/H_2 = (H/H_2)^2$,

$$\alpha \approx \left(1 + \varepsilon \frac{H_2}{H}\right) \frac{H_2}{H_1} \quad \text{or} \quad \alpha \approx -\left(1 + \varepsilon \frac{H_1}{H}\right).$$

These solutions correspond to a barotropic mode and a baroclinic mode.

Barotropic mode:

$$\alpha = (1 + \varepsilon H_2/H)H_2/H_1, \quad c^2 = gH, \quad p = g\eta H/H_1 + \varepsilon g h H_2/H_1. \quad (3.39)$$

Baroclinic mode:

$$\alpha = -(1 + \varepsilon H_1/H), \quad c^2 = g H_1 H_2 \varepsilon/H, \quad p = -g\varepsilon h + g\varepsilon \eta H_2/H. \quad (3.40)$$

Note that when the motion is 'barotropic' the baroclinic mode pressure is zero. This implies that

$$g\varepsilon h = g\varepsilon\eta \, H_2/H.$$

Hence barotropic motion has $h \sim \eta$ and

$$p \text{ (barotropic)} \approx g\eta H/H_1. \quad (3.41)$$

Substitute this result, $c^2 = gH$ and $\alpha \approx H_2/H_1$ into the equations (3.37a,b,c) to get, after multiplying through by H_1/H,

$$u_{0t} - fv_0 = -g\eta_x + \tau^x/\rho_1 H \quad (3.42a)$$

$$v_{0t} + fu_0 = -g\eta_y + \tau^y/\rho_1 H \quad (3.42b)$$

$$\eta_t + H(u_{0x} + v_{0y}) = 0 \quad (3.42c)$$

where

$$\mathbf{u}_0 = (H_1 \mathbf{u}^{(1)} + H_2 \mathbf{u}^{(2)})/H. \quad (3.43)$$

Thus the equations for the barotropic mode are equivalent to (3.42) and (3.43), the equations for depth-averaged flow given that the pressure is independent of depth. We note that the pressure is essentially independent of depth for the barotropic mode because upper and lower layers have nearly the same pressure perturbations for ε small and $h \sim \eta$ (see (3.36a) and (3.36c)). We could have anticipated that the barotropic mode would be one of the modes because the result (3.24) is true for general N^2 with density difference over the ocean depth much smaller than typical density.

Similarly, when the motion is 'baroclinic,' the barotropic mode pressure is zero and so from (3.39) $\eta \sim \varepsilon h$. Therefore from (3.40)

$$p \text{ (baroclinic)} \approx -g\varepsilon h.$$

The baroclinic mode has $\alpha \approx -1$ and the baroclinic mode results are

$$\mathbf{u}_1 = \mathbf{u}^{(1)} - \mathbf{u}^{(2)}, \quad p_1 = -g\varepsilon h, \quad c_1^2 = g\varepsilon H_1 H_2/H, \quad b_1 = 1/H_1. \quad (3.44)$$

The reason for spending so much time on the two layer case is that in certain places (e.g., near the equator) the ocean has stratification resembling that of a two-layer fluid. For example, in Fig. 3.2 the strong thermocline approximates an interface between two layers of much more slowly varying density with $H_1 << H$ (typically $H_1 \leq 100$ m and $H \sim 4$ km). The equatorial response is dominated by the baroclinic mode because the baroclinic

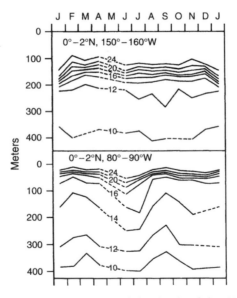

Fig. 3.2 Variation during a typical year of the depth of the 10–24°C isotherms at 2°C intervals in the central Pacific (top) and eastern Pacific (bottom), determined by averaging the bathythermographic observations in 2° latitude by 10° longitude boxes. Dashed lines indicate missing data. Note that the stratification is qualitatively like the two-layer ocean model in that temperature (and hence density) changes rapidly over a small depth. It differs from the two-layer model in that in the real ocean the 'lower layer' is stratified and the rapid change in density occurs over a depth that is not small compared to the depth of the upper layer. (From Meyers 1979.)

forcing enters as $\tau/\rho_1 H_1$ (see (3.25), (3.26), (3.28a) and (3.44)) whereas the barotropic forcing enters as the much smaller vector $\tau/\rho_1 H$ (see (3.42a) and (3.42b)). We therefore expect $p_0 \approx 0$ and hence

$$\eta = -\varepsilon \frac{H_2}{H} h \approx -\varepsilon h. \tag{3.45}$$

That this is approximately true is shown (see Fig. 3.3) by the results of Meyers (1979).

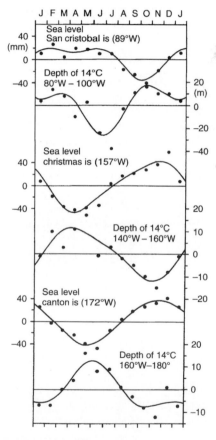

Fig. 3.3 Annual variation in sea level (mm) measured at equatorial islands and upward displacement (m) of the 14°C isotherm in the nearby ocean. Dots give average monthly values and the smooth curves are a least–squares best fit to the first two harmonics (frequencies $2\pi/12$ months and $2\pi/6$ months). Consistent with (3.45), upward thermocline displacement (as modeled by the 14°C isotherm displacement) and sea level are generally 180° out of phase with $h \gg \eta$. (From Meyers 1979.)

Equation (3.45) shows that the thermocline displacement is $\varepsilon^{-1} \approx$ 200—300 times the sea level displacement and is of opposite sign. Physically, because the upper layer is directly driven by the windstress and is thin compared to the lower layer, we expect the currents and associated pressure gradients to be much stronger there than in the lower layer. Negligible pressure gradients in the lower layer implies that if the sea level is (say) higher than normal, the upper layer must be much thicker than normal so that the total pressure in the lower layer does not change – the extra pressure $\rho_1 g \eta$ (with $\eta > 0$) is compensated by reduced pressure $(\rho_2 - \rho_1) g h$ (with $h < 0$) due to there being extra lighter fluid over the vertical distance h. Since $\rho_2 - \rho_1 << \rho_1$, h must be much larger than η for this pressure compensation.

3.4. EQUATORIAL WAVES

Fundamental to understanding equatorial variability is an understanding of the possible linear equatorial waves. Our analysis will be based on the modal equations (3.25)–(3.27) under the equatorial β-plane approximation $f = \beta y$. We put $\tau^x = \tau^y = 0$ because we seek the free waves. The equations are thus (dropping the n subscript for notational convenience):

$$u_t - \beta y v = -p_x \tag{3.46}$$

$$v_t + \beta y u = -p_y \tag{3.47}$$

$$p_t / c^2 + u_x + v_y = 0. \tag{3.48}$$

3.4.1. The equatorial Kelvin wave

One solution to (3.46)–(3.48) has north–south velocity identically zero and therefore satisfies the equations

$$u_t = -p_x \tag{3.49}$$

$$\beta y u = -p_y \tag{3.50}$$

$$p_t / c^2 + u_x = 0. \tag{3.51}$$

The general solution to this system that is finite for large distances from the equator is

$$u = p/c = G(x - ct) \; \exp(-y^2 \beta / 2c) \tag{3.52}$$

where G is a general (differentiable) function. This wave solution represents an equatorial Kelvin wave. It is trapped near the equator with a decay scale $(2c/\beta)^{1/2}$ and travels eastward at speed c since the amplitude function G does not change from $G(x_0)$ if $(x - x_0)/t = c$. This wave is nondispersive — if one looks for a solution of the form $\exp(ikx - i\omega t)\exp(-y^2\beta/2c)$ one obtains the form (3.52) with

$$G = \exp(ikx - i\omega t) \text{ and } \omega/k = c \text{ (constant).} \qquad (3.53)$$

Physically, the equatorial Kelvin wave is a long gravity wave trapped to the equator by rotation. It is a gravity wave because (3.49) and (3.51) are the equations for a long gravity wave traveling in the zonal direction. For the barotropic mode, $c = (gH)^{1/2}$, the speed of a long gravity wave in constant density water. The much slower baroclinic mode speeds depend on N and therefore gravity and vertical gradients of density.

Often the ocean response is dominated by the first few baroclinic modes. The first baroclinic mode travels eastward at a speed $c \approx 2 - 3\,\mathrm{m\,s^{-1}}$ and is trapped near the equator with a decay scale $(2c/\beta)^{1/2} \approx 470\,\mathrm{km}$ for $c = 2.5\,\mathrm{m\,s^{-1}}$ and $\beta = 2.3 \times 10^{-11}\,\mathrm{m^{-1}\,s^{-1}}$. Higher order baroclinic modes (smaller c) are even more tightly trapped to the equator.

Why must the equatorial Kelvin wave travel eastward and not westward along the equator? Because the equatorial Kelvin wave is a gravity wave, currents underneath a 'crest' (or in a region of high pressure perturbation) must be in the direction of wave propagation while underneath a 'trough' (low pressure perturbation) currents are in the direction opposite to wave propagation. We can understand this using Fig. 3.4. In order for the crest C

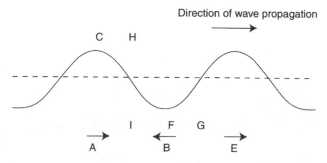

Fig. 3.4 Propagation of a gravity wave. In order for the crest C to be at H and the trough F to be at G one quarter of a period later, currents at A and B must now be converging toward I and at B and E diverging from G. Therefore, in a gravity wave, currents are in the direction of wave propagation underneath a crest and opposite to the direction of wave propagation underneath a trough.

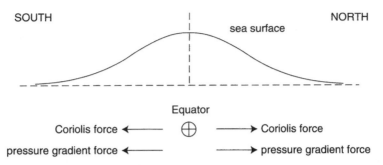

Fig. 3.5 North–south cross-section through a crest of a westward propagating equatorial Kelvin wave. Since the wave propagates westward, the current underneath the wave crest is westward and the Coriolis force, being to the right of the current in the Northern Hemisphere and to the left of the current in the Southern Hemisphere, is as shown, i.e., in the same direction as the pressure gradient force trying to flatten the sea level. Since both forces are in the same direction and therefore cannot balance, a westward propagating equatorial Kelvin wave is not viable.

to move in the direction of wave propagation to H one quarter of a period later, currents at A and B converge on I as shown. These current directions are consistent with the above rule. Similarly, in order for the trough at F to propagate to G one quarter of a period later, currents diverge at G as shown. This is also consistent with the above rule. Suppose now we try to make the equatorial Kelvin wave propagate westward. Consider a north–south cross-section of a crest (see Fig. 3.5). Since the wave is traveling westward, the current must be into the paper underneath a crest and consequently the pressure gradient and Coriolis forces cannot balance. Thus westward propagation is impossible. Analogous diagrams for eastward propagation for both crest and trough cases show that pressure gradient and Coriolis forces can balance.

3.4.2. Other equatorial waves

To examine other possible equatorial waves, it is convenient to non-dimensionalize (3.46)–(3.48) by writing

$$u = cu', v = cv', p = c^2 p', x = \sqrt{\frac{c}{\beta}} x', y = \sqrt{\frac{c}{\beta}} y' \text{ and } t = \frac{t'}{\sqrt{\beta c}}.$$

Then in terms of non-dimensional (dashed) variables, (3.46)–(3.48) can be written as follows:

$$\frac{\partial u'}{\partial t'} - y' v' = -\frac{\partial p'}{\partial x'} \tag{3.54}$$

$$\frac{\partial v'}{\partial t'} + y'u' = -\frac{\partial p'}{\partial y'}$$ (3.55)

$$\frac{\partial p'}{\partial t'} + \frac{\partial u'}{\partial x'} + \frac{\partial v'}{\partial y'} = 0.$$ (3.56)

For notational convenience we will drop the dashes from the non-dimensional variables in what follows.

By differentiating (3.54) with respect to t and (3.56) with respect to x, non-dimensional pressure p may be eliminated between (3.54) and (3.56) to give

$$u_{tt} - u_{xx} = yv_t + v_{xy}.$$ (3.57)

Similarly, by differentiating (3.55) with respect to t and (3.56) with respect to y, p may be eliminated between (3.55) and (3.56) to give

$$v_{tt} - v_{yy} = -yu_t + u_{yx}.$$ (3.58)

It is also convenient to eliminate u from (3.57) and (3.58) to obtain an equation in terms of v alone. Applying $(\partial^2/\partial t^2 - \partial^2/\partial x^2)$ to (3.58) gives, using (3.57) and one integration with respect to time,

$$v_{xxt} + v_{yyt} + v_x - y^2 v_t - v_{ttt} = 0.$$ (3.59)

Separable wave solutions of (3.59) are

$$v = \exp(ikx - i\omega t)\, \psi(y)$$ (3.60)

with

$$\psi_{yy} + \left(\omega^2 - k^2 - \frac{k}{\omega} - y^2 \right) \psi = 0.$$

We need $\psi \to 0$ as $|y| \to \infty$ since the solutions must be bounded. This is possible only when

$$\omega^2 - k^2 - \frac{k}{\omega} = 2m+1 \qquad m = 0, 1, \dots$$

or

$$k^2 + \frac{k}{\omega} - \omega^2 + (2m+1) = 0.$$ (3.61)

Then

$$v = \exp(ikx - i\omega t)\, \psi_m(y) \qquad (3.62)$$

where the Hermite function $\psi_m(y)$ satisfies

$$\frac{d^2\psi_m}{dy^2} + (2m + 1 - y^2)\,\psi_m = 0. \qquad (3.63)$$

Equation (3.63) indicates that $\psi_m(y)$ is oscillatory for $|y| \le (2m + 1)^{1/2}$ and monotonic (decaying) for $|y| > (2m + 1)^{1/2}$. $|y| = (2m + 1)^{1/2}$ is the 'turning latitude' for the mode being considered.

The Hermite function $\psi_m(y)$ is of the form

$$\psi_m(y) = \frac{\exp(-y^2/2)H_m(y)}{\sqrt{2^m \cdot m! \sqrt{\pi}}} \qquad (3.64a)$$

where $H_m(y)$ is the mth Hermite polynomial. The first few Hermite polynomials are

$$H_0(y) = 1, \quad H_1(y) = 2y, \quad H_2(y) = 4y^2 - 2, \quad H_3(y) = 8y^3 - 12y,$$
$$H_4(y) = 16y^4 - 48y^2 + 12, \quad H_5(y) = 32y^5 - 160y^3 + 120y.$$
$$(3.64b)$$

Notice that ψ_m is an even function of y when m is even and is an odd function of y when m is odd. Hermite functions satisfy the orthogonality property

$$\int_{-\infty}^{\infty} \psi_m \psi_n \, dy = \begin{matrix} 0 \\ 1 \end{matrix} \quad \begin{matrix} (m \ne n) \\ (m = n) \end{matrix}. \qquad (3.65)$$

Solving (3.61) for k gives

$$k = \frac{-1}{2\omega} \pm \sqrt{\omega^2 + \frac{1}{4\omega^2} - (2m + 1)}. \qquad (3.66)$$

Two main cases, $m = 0$ and $m \ne 0$, arise.

(i) The special case $m = 0$
    When $m = 0$, the two roots for k are

$$k = \omega - \frac{1}{\omega} \quad \text{and} \quad k = -\omega. \qquad (3.67)$$

The solution with

$$v = \exp\left\{i\left(\omega - \frac{1}{\omega}\right)x - i\omega t\right\}\psi_0(y) \qquad (3.68)$$

is called a Yanai wave, or 'mixed' Rossby gravity wave. For large ω one has

$$k \approx \omega \qquad (3.69a)$$

which is the non-dimensional equatorial Kelvin wave (long *gravity* wave) dispersion relation. For small ω

$$k \approx -\frac{1}{\omega} \qquad (3.69b)$$

which is the high-zonal wave number limit for the *Rossby* wave (see (3.79)). The cross-over point, $\omega = 1$, corresponds to a dimensional period $T = 2\pi/\sqrt{\beta c}$ (≈ 9.6 days for the first baroclinic mode using $c = 2.5\,\mathrm{m\,s^{-1}}$ and $\beta = 2.3 \times 10^{-11}\,\mathrm{m^{-1}\,s^{-1}}$).

The u and p fields corresponding to (3.68) are

$$u = p = \frac{i\omega}{\sqrt{2}}\exp\left\{i\left[\left(\omega - \frac{1}{\omega}\right)x - \omega t\right]\right\}\psi_1(y). \qquad (3.70)$$

This result can be deduced using certain recurrence relations (see later in this section).

The other root, $k = -\omega$, corresponds to a westward propagating Kelvin wave. Since v is proportional to $\psi_0(y)$ which approaches zero for large $|y|$, at large distances from the equator, by (3.58)

$$u_y \to yu,$$

i.e., u is proportional to $\exp(y^2/2)$. Thus u becomes unbounded for large $|y|$ and hence the solution with $k = -\omega$ must be rejected.

(ii) $m \geq 1$

For $m \geq 1$, both roots of (3.66) lead to acceptable wave fields. There are two classes of waves: first, a relatively high-frequency inertia-gravity wave set (i.e., gravity waves modified by the earth's rotation) and second, a low-frequency class of Rossby waves. Groves (1955) first

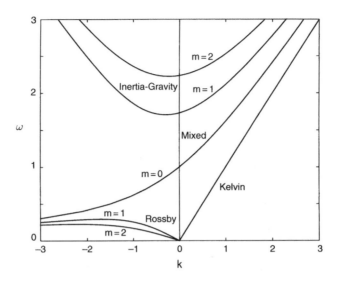

Fig. 3.6 The dispersion relation for equatorial waves.

plotted the dispersion diagram (Fig. 3.6) for these two wave classes and also the Kelvin and mixed Rossby gravity waves.

If (3.61) is differentiated with respect to k, we find that the group velocity in the x direction is

$$c_g = \frac{\partial \omega}{\partial k} = \frac{2k\omega + 1}{2\omega^2 + k/\omega}. \tag{3.71}$$

Thus the frequency extrema occur when $2k\omega = -1$. One can find these extrema by putting $k = -1/2\omega$ in (3.61). Minima occur at

$$\omega = \left(\frac{m+1}{2}\right)^{1/2} + \left(\frac{m}{2}\right)^{1/2} \tag{3.72a}$$

for the high-frequency inertia-gravity waves and maxima occur at

$$\omega = \left(\frac{m+1}{2}\right)^{1/2} - \left(\frac{m}{2}\right)^{1/2} \tag{3.72b}$$

for the low-frequency Rossby waves.

The u and p fields

In establishing the u and p fields we make use of the following recurrence relations for the Hermite functions $\psi_m(y)$ defined by (3.64):

$$\frac{d\psi_m}{dy} = -\frac{1}{\sqrt{2}}(m+1)^{1/2}\,\psi_{m+1} + \frac{1}{\sqrt{2}}m^{1/2}\,\psi_{m-1} \tag{3.73a}$$

$$y\psi_m = \frac{1}{\sqrt{2}}(m)^{1/2}\,\psi_{m-1} + \frac{1}{\sqrt{2}}(m+1)^{1/2}\,\psi_{m+1}. \tag{3.73b}$$

The u field   From (3.57) and (3.60)

$$\left(k^2 - \omega^2\right)u = e^{ikx - i\omega t}\left(-i\omega y\psi_m + ik\frac{d\psi_m}{dy}\right). \tag{3.74}$$

Using the recurrence relations for $y\psi_m$ and $d\psi_m/dy$ we thus have

$$u = \left(e^{ikx-i\omega t}\right)\frac{i}{2}\left\{\frac{\sqrt{2m}\psi_{m-1}}{(k+\omega)} + \frac{\sqrt{2(m+1)}\psi_{m+1}}{(\omega-k)}\right\}. \tag{3.75}$$

The p field   Differentiating (3.54) with respect to x, (3.56) with respect to t and subtracting the results gives

$$p_{xx} - p_{tt} = yv_x + v_{yt}. \tag{3.76}$$

Similarly to the u field, it follows from (3.60) and the recurrence relations (3.73) that

$$p = e^{ikx-i\omega t}\left(\frac{i}{2}\right)\left\{\frac{\sqrt{2(m+1)}\psi_{m+1}}{\omega-k} - \frac{\sqrt{2m}\psi_{m-1}}{\omega+k}\right\}. \tag{3.77}$$

Note that for the special case when $m = 0$ the recurrence relations are

$$\frac{d\psi_0}{dy} = -\frac{1}{\sqrt{2}}\psi_1 \tag{3.78a}$$

$$y\psi_0 = \frac{1}{\sqrt{2}}\psi_1 \tag{3.78b}$$

and a similar analysis to the above gives the Yanai wave p and u fields in (3.70).

3.4.3. Low-frequency equatorial waves

Since our interest is in ENSO, we will concentrate on the low-frequency equatorial waves. Typical ENSO frequencies are of order $2\pi/3$ years. For a first baroclinic mode ($c = 2.7\,\mathrm{m\,s^{-1}}$), the corresponding non-dimensional frequency is 0.008 and at this frequency Fig. 3.6 indicates that only very long ($k \approx 0$) equatorial Kelvin waves, very long equatorial Rossby waves or very short ($|k| >> 1$) Rossby and Yanai waves are possible. It follows from (3.66) that at low frequencies the Rossby wave dispersion relationship is

$$k \approx \frac{-1}{2\omega} \pm \frac{1}{2\omega} \left[1 - (2m+1)\,4\omega^2\right]^{1/2},$$

i.e., by the binomial expansion,

$$k \approx \frac{-1}{\omega} \quad \text{or} \quad -(2m+1)\,\omega. \tag{3.79}$$

Thus the large-zonal scale ($k \approx 0$) Rossby waves are non–dispersive (since $\omega/k = -1/(2m+1)$) and the short-zonal scale Rossby waves have the same dispersion relation ($k \equiv -1/\omega$) as the low-frequency Yanai wave (see (3.69b)).

From (3.75), (3.77) and $k = -(2m+1)\omega$, the large-zonal scale, low-frequency Rossby wave u and p fields have the form

$$u = 2^{-3/2}\omega^{-1}\mathrm{i}\exp[-\mathrm{i}\omega((2m+1)x+t)]\left\{\frac{\psi_{m+1}}{(m+1)^{1/2}} - \frac{\psi_{m-1}}{(m)^{1/2}}\right\} \tag{3.80}$$

and

$$p = 2^{-3/2}\omega^{-1}\mathrm{i}\exp\left[-\mathrm{i}\omega((2m+1)x+t)\right]\left\{\frac{\psi_{m+1}}{(m+1)^{1/2}} + \frac{\psi_{m-1}}{(m)^{1/2}}\right\}. \tag{3.81}$$

The large-zonal scale v Rossby wave field is of order ω compared to the u field (see (3.80) and (3.62)) and therefore the horizontal velocity is nearly zonal. Note that x and t in the p and u fields only occur in the combination $(2m+1)x+t$ so that if we sum over all frequencies in a general low-frequency, large-scale Rossby wave disturbance we may write

$$u = Q((2m+1)\,x+t)\left\{\frac{\psi_{m+1}}{(m+1)^{1/2}} - \frac{\psi_{m-1}}{(m)^{1/2}}\right\} \tag{3.82}$$

and

$$p = Q((2m+1)x+t) \left\{ \frac{\Psi_{m+1}}{(m+1)^{1/2}} + \frac{\Psi_{m-1}}{(m)^{1/2}} \right\} \qquad (3.83)$$

where Q is an amplitude function.

3.4.4. Energy flux and group velocity

Define the non–dimensional energy flux F and non–dimensional energy density E by

$$F = \int_{-\infty}^{\infty} pu \, dy \qquad (3.84)$$

and

$$E = \frac{1}{2} \int_{-\infty}^{\infty} (u^2 + p^2) dy. \qquad (3.85)$$

Normally the energy density would include a term $v^2/2$ in the integrand since this contributes to the non–dimensional kinetic energy $(u^2 + v^2)/2$ in the energy density. However, here we are concerned with equatorial Kelvin or long equatorial Rossby waves for which v is either identically zero or negligible.

For the equatorial Kelvin wave it follows from (3.52) and the non–dimensionalization of dimensional p by c^2 and dimensional u by c that

$$p = u \qquad (3.86)$$

and hence, by (3.84) and (3.85) that

$$F = c_g E \qquad (3.87)$$

since the non–dimensional group velocity c_g for the equatorial Kelvin wave is one. The relationship (3.87) also holds for the long equatorial Rossby waves. This follows from (3.82), (3.83), the orthogonality of the Hermite functions and the long equatorial Rossby wave group velocity result (see (3.79))

$$c_g = \frac{\partial \omega}{\partial k} = \frac{-1}{(2m+1)}. \qquad (3.88)$$

Since E is always positive, (3.87) indicates that F has the same sign as c_g. Since c_g is negative for the long equatorial Rossby waves, for these waves p and u are, on average, of opposite sign. This result will prove useful later in our discussion of ENSO dynamics.

 ## REFERENCES

Clarke, A. J., and S. Van Gorder, 1986: A method for estimating wind-driven frictional, time-dependent, stratified shelf and slope water flow. *J. Phys. Oceanogr.*, **16**, 1013–1028.

Groves, G. W., 1955: Day to day variation of sea level. PhD Thesis, Scripps Institution of Oceanography, University of California, Los Angeles, 88 pp.

Kundu, P. K., 1990: *Fluid Mechanics*. Academic Press, New York, 638 pp.

Meyers, G., 1979: Annual variation of the slope of the 14°C isotherm along the equator in the Pacific Ocean. *J. Phys. Oceanogr.*, **9**, 885–891.

CHAPTER FOUR

EQUATORIAL WAVE REFLECTION FROM PACIFIC OCEAN BOUNDARIES

Contents

4.1. OVERVIEW

For thousands of kilometers along the western coastlines of North and South America zooplankton and many bird and fish populations are sensitive to ocean fluctuations associated with ENSO. ENSO also affects marine populations in the western Pacific. To understand these fluctuations and also the basic dynamics of ENSO, it is necessary to study low-frequency equatorial ocean wave reflection. We will discuss this reflection for the western Pacific (Section 4.2) and eastern Pacific (Section 4.3).

4.2. Wave Reflection at the Western Pacific Boundary

Since the energy of waves travels at the group velocity, waves incident to a boundary have group velocity towards the boundary while waves reflected from the boundary will have group velocity directed away from that boundary. For equatorial waves the group velocity

$$c_g = \frac{\partial \omega}{\partial k} \tag{4.1}$$

is eastward when c_g is positive and westward when c_g is negative. During ENSO, low-frequency large-zonal scale wind forcing generates large-zonal scale ($k \approx 0$), low-frequency ($\omega \approx 0$) waves. Therefore, the ENSO ocean response will be in terms of those waves near the origin of the (ω, k) dispersion diagram (Fig. 3.6), viz., the equatorial Kelvin wave, with eastward group velocity ($\partial \omega / \partial k > 0$) and the westward group velocity Rossby waves ($\partial \omega / \partial k < 0$). Since the only large-scale waves with westward group velocity are the Rossby waves, for us these are the only relevant waves incident to a western boundary.

4.2.1. Reflection from a solid meridional western boundary

The western boundary of the Pacific is irregular and has many gaps. However, before we can understand the reflection from this complicated boundary, we begin with the simpler case of the reflection of Rossby waves from a solid (non-gappy) meridional boundary.

Moore (1968) first considered the reflection of equatorial waves from ocean boundaries. He showed that the reflection of an incoming mode M equatorial Rossby wave of frequency ω from a north–south solid boundary could be found in closed form as a finite sum of short eastward group velocity equatorial Rossby waves ($m \leq M$) and a short-wavelength Yanai wave (if M is even) or an equatorial Kelvin wave (if M is odd). Moore obtained his solution by first writing the zonal velocity u as a sum of the zonal velocity u_{inc} of the incoming Rossby wave and a sum of the reflected wave zonal velocity fields u_{ref} and then using the boundary condition that the total zonal velocity must vanish at the meridional wall to obtain the amplitudes of the reflected waves. This solution is documented in Clarke (1983).

4.2.2. Low-frequency reflection from a solid non-meridional western boundary

At low frequencies Cane and Gent (1984) generalized Moore's solution to a solid western ocean boundary of arbitrary shape. Their results, derived below, provide a simple physical explanation of low-frequency solid western boundary reflection.

As noted earlier, only westward group velocity ($c_g < 0$) waves can be incident to a western boundary and those reflect as low-frequency eastward group velocity waves. The latter are of two types: the large-zonal scale equatorial Kelvin wave and the very short eastward group velocity Rossby waves and Yanai wave. For the short waves $k = -1/\omega$ (see (3.69b) and (3.79)). Under this approximation, it follows from (3.68), (3.70) and the recurrence relation (3.78a) that the low-frequency Yanai wave velocity field satisfies

$$u_x + v_y = 0.$$

A similar result holds for the short Rossby wave velocity field using $k = -1/\omega$, (3.62), (3.75) and the recurrence relation (3.73a). Thus for all the short waves at all low frequencies the horizontal velocity field is non-divergent:

$$\nabla \cdot \mathbf{u}_B = 0 \qquad (4.2)$$

where $\mathbf{u}_B$ refers to the velocity field of all the short waves. We have used the subscript B (for boundary) because these waves have very small group velocity ($\partial \omega / \partial k = \omega^2$) and for even a small dissipation will be trapped near the boundary. In the real world non-linear effects will also be important near the boundary but when these are included in a $1\frac{1}{2}$ layer ocean model (Clarke 1991), similar results to those derived here are valid.

Integration of (4.2) over the region bounded by the irregular solid western boundary, northern ($y = y_N$) and southern latitudes ($y = y_S$) and a meridian $x = x_0$ (see Fig. 4.1) gives, by the divergence theorem,

$$\int_{\substack{y=y_S \\ n=0}}^{y=y_N} (-\mathbf{e}_n) \cdot \mathbf{u}_B \, ds + \int_A^C v_B \, dx + \int_C^D u_B(-dy) + \int_D^B (-v_B)(-dx) = 0$$

$$(4.3)$$

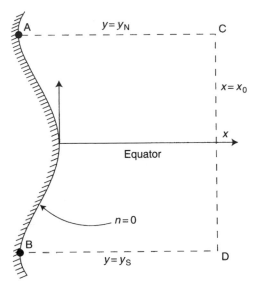

Fig. 4.1 Integration region near the solid irregular western ocean boundary. The coordinate n is the normal distance from the irregular boundary.

where $\mathbf{e}_n$ is the unit normal vector pointing seaward from the western boundary $n = 0$. Since $\mathbf{u}_B$ is a boundary layer velocity, the third integral on the left-hand side of (4.3) vanishes for large enough x_0. Also, since the incoming energy is assumed to have effectively finite meridional extent, $\mathbf{u}_B$ will also have finite meridional extent. Taking y_N and $|y_S|$ large enough implies that the second and fourth integrals on the left-hand side of (4.3) also vanish and thus

$$\int_{\substack{y_S \\ n=0}}^{y_N} \mathbf{e}_n \cdot \mathbf{u}_B \, ds = 0 \ . \tag{4.4}$$

To formulate the western boundary condition in terms of the large-scale flow, write the total horizontal velocity $\mathbf{u}$ as a sum of $\mathbf{u}_B$ and the large-scale velocity field $\mathbf{u}_L$. Then the condition of no normal flow through the western boundary implies that

$$\mathbf{e}_n \cdot \mathbf{u}_L + \mathbf{e}_n \cdot \mathbf{u}_B = 0 \qquad \text{on} \ \ n = 0$$

and hence, from (4.4), that

$$\int_{\substack{y_S \\ n=0}}^{y_N} \mathbf{e}_n \cdot \mathbf{u}_L \, ds = 0. \tag{4.5}$$

Physically, (4.5) states that the net large-scale mass flow on to the western boundary must be zero. The boundary layer currents $\mathbf{u}_B$ thus serve to take the large-scale incoming Rossby mass flow and provide all of it to the mass flow of the outgoing equatorial Kelvin wave (Cane and Sarachik 1977; Cane and Gent 1984).

For a western boundary defined by $x = b(y)$, the unit normal to the boundary is

$$\mathbf{e}_n = \nabla(x - b)/|\nabla(x - b)| \qquad (4.6)$$

and so (4.5) can be written as

$$\int_{y_S}^{y_N} \left[u_L(b(y), y) - \frac{db}{dy} v_L(b(y), y) \right] dy = 0 \qquad (4.7)$$

where we have used

$$ds = dy \left(1 + \left(\frac{db}{dy} \right)^2 \right)^{1/2}.$$

As noted earlier in Section 3.4.3, at low frequencies the large-scale equatorial waves have negligible v velocity and so only the $u(b(y), y)$ term matters in (4.7). But

$$u(b(y), y) = u(0, y) + bu_x(0, y) + \cdots \qquad (4.8)$$

and u_x is proportional to ω for the large-scale equatorial waves. Hence at low enough frequencies, (4.7) reduces to

$$\int_{y_S}^{y_N} u(0, y) dy = 0, \qquad (4.9)$$

i.e., the boundary is dynamically like a meridional wall. We will exploit this idea in the next section when we consider reflection from the gappy western Pacific boundary at ENSO frequencies.

4.2.3. ENSO frequency reflection from the gappy western Pacific boundary

The real western Pacific boundary is not solid but discontinuous, extending east–west over 4000 km and consisting of several irregularly shaped land masses (Fig. 4.2). Just how well does this gappy irregular

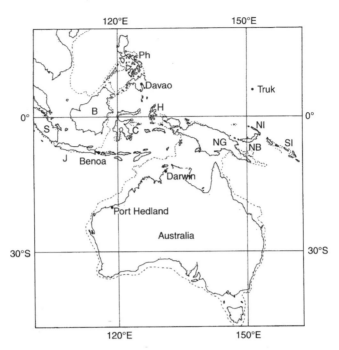

Fig. 4.2 The gappy, irregular western Pacific boundary region. The dashed line is the 200 m isobath. SI, Solomon Islands; NB, New Britain; NI, New Ireland; NG, New Guinea; H, Halmahera; C, Celebes; Ph, Philippines; B, Borneo; J, Java; S, Sumatra. (From Clarke 1991).

boundary reflect ENSO energy? Where does the reflection mainly occur? Does the ENSO Pacific Ocean signal get into the Indian Ocean via an ocean path? The first theories to discuss low-frequency reflection at the gappy western Pacific boundary were those by du Penhoat and Cane (1991) and Clarke (1991). We follow Clarke's treatment. In that paper detailed mathematical derivations are given; here we present the essential ideas.

In addressing the questions raised above, we will utilize the result that, essentially because low-frequency large-scale equatorial waves have such a large east–west scale, land masses forming the western tropical Pacific boundary can each be treated dynamically as being infinitesimally thin east–west and meridional. Thus it can be shown that at low frequencies the western tropical Pacific boundary consists dynamically of several thin meridional islands (Fig. 4.3). Actually, to simplify the problem mathematically, the *whole* western boundary region, not just each island, must be dynamically thin, i.e., the phase of the dominant waves in the reflection

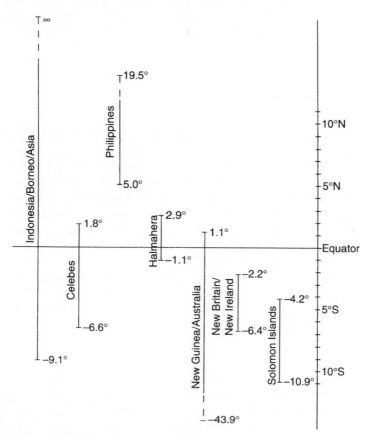

Fig. 4.3 Thin island approximations to the western Pacific boundary. The northern and southern latitudes were based on the northern and southern limits of the 200 m isobath for each island. (From Clarke 1991).

must vary negligibly across the region of width $\Delta x \approx 4900$ km. Expanding p and u about a longitude $x = 0$ centrally located in the western boundary region gives, for example, for u

$$u(x) = u(0) + xu_x(0) + \cdots . \tag{4.10}$$

Thus the motion is independent of x if

$$k\Delta x/2 \ll 1 \tag{4.11}$$

where k is the largest wave number of the dominant waves involved in the reflection/transmission process. Since the equatorial Kelvin and first mode

Rossby waves dominate, the largest wave number is, in dimensional form, $k = 3\omega/c$. Substitution into (4.11) using $c = 2.7\,\mathrm{m\,s}^{-1}$, $\Delta x = 4900\,\mathrm{km}$ and $\omega = 2\pi/3$ years (a typical frequency for ENSO) gives

$$3\omega \Delta x/2c = 0.18 \qquad (4.12)$$

which, as a maximum error, is small. Thus, dynamically, the western boundary of the Pacific is thin and looks like the seven thin islands shown in the figure.

The $m = 1$ Rossby wave is dominant among the Rossby waves taking part in the Pacific western boundary reflection (Clarke 1982; Battisti 1988; Clarke 1991; Kessler 1991; Wakata and Sarachik 1991). Consider this Rossby wave approaching the gappy western boundary. The Rossby wave u field shown (Fig. 4.4) suggests that the wave is incompletely blocked by each island except the last one (Indonesia/Borneo/Asia) which essentially acts like an infinite meridional wall. Calculations verify that the reflection mainly occurs at this last 'island' and that essentially no energy from the Pacific gets past Indonesia/Borneo/Asia. The equatorial Kelvin wave reflected back into the Pacific has an amplitude of about 83% of that which would be reflected from a solid meridional wall.

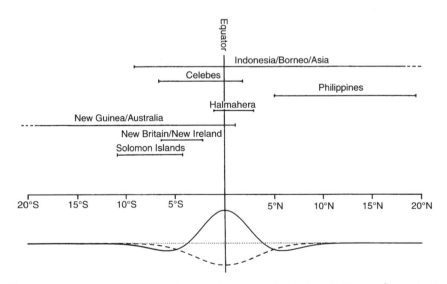

Fig. 4.4 First meridional mode first vertical mode ($c = 2.7\,\mathrm{m\,s}^{-1}$) Rossby wave (solid curve) and Kelvin wave (dashed curve) non–dimensional u fields compared with the latitudinal extent of the islands. Only Indonesia/Borneo/Asia extends beyond the scale of the u fields. The dotted line is the zero amplitude line. (From Clarke 1991).

Although the Indonesia/Borneo/Asia 'island' blocks the Pacific equatorial ocean ENSO variability, the Australia/New Guinea land mass does not since this island does not extend completely across the u field of the main ($m = 1$) equatorial Rossby wave (see Fig. 4.4). This implies that ENSO energy of Pacific equatorial origin should occur along Australia's western coast. Sea level observations (Pariwono et al. 1986; Clarke 1991) confirm this. On the other hand, negligible equatorial Indian Ocean signal gets to the Darwin coast because the only low-frequency large-scale wave with eastward group velocity is the equatorial Kelvin wave and the Indian Ocean equatorial Kelvin wave is largely blocked by the Indonesia/Borneo/Asia land mass which extends right across the equatorial wave guide. Thus the sea level along Australia's western coast is of western equatorial Pacific origin and can be used to estimate the reflected equatorial Kelvin wave at the western Pacific boundary (Li and Clarke 1994; Mantua and Battisti 1994).

The ENSO sea level signal and associated currents and thermocline displacement along Australia's western and southern coastlines have a profound influence on the ecosystems there. This will be discussed in Sections 12.5 and 12.6.

4.2.4. Interannual surface flow between the Pacific and Indian Oceans

The gappy western Pacific boundary allows an interannual exchange of mass between the Pacific and Indian Oceans and this mass exchange is related to the reflection dynamics of the equatorial waves in both oceans. Here we describe the likely dynamics of this interannual flow.

Both physically (see Fig. 4.2) and dynamically (Clarke 1991) the New Guinea/Australia and Indonesia/Borneo/Asia land masses dominate the reflection of low-frequency energy in the western equatorial Pacific. The ENSO Pacific Ocean signal on Australia's western coastline, according to theory, should propagate westwards as large-scale Rossby waves and introduce the Pacific ENSO signal into the Indian Ocean interior south of about 9°S, the latitude of the southern boundary of the Indonesia/Borneo/Asia land mass. At ENSO frequencies the flow is nearly geostrophic and so, to avoid flow into the solid western Australian coast, the pressure (and hence sea level) is nearly spatially constant along the coast. Since the westward propagating Indian Ocean Rossby waves are of large-zonal scale at ENSO frequencies, Indian Ocean ENSO pressure fluctuations should be nearly spatially constant off Australia's northwest coast. Satellite estimated sea level heights (Feng et al. 2003; Clarke and Li 2004) confirm this. North of 9°S and west of Indonesia/Borneo/Asia there is no ENSO signal since equatorial Pacific Rossby waves are blocked by Indonesia/Borneo/Asia. Thus, according to theory, immediately south of Indonesia/Borneo/Asia there is a sharp discontinuity in sea level and pressure extending zonally west of the tip of this island (see Fig. 4.5 with p_S non-zero and $p_N = 0$). Associated with this sharp pressure gradient is, by geostrophy, a zonal current carrying water between the Pacific and Indian Oceans. During a warm ENSO episode, surface equatorial Pacific winds are anomalously westerly (see Chapter 2), anomalous sea level is tilted up toward the east and consequently sea level in the western equatorial Pacific and along Australia's western coast is lower than normal. Lower sea level

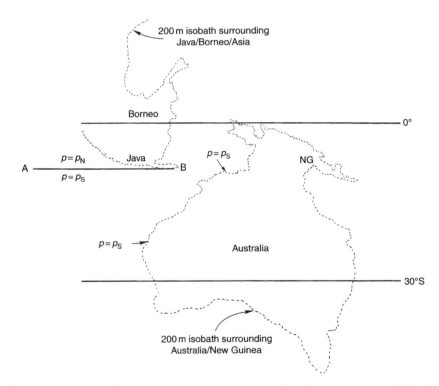

Fig. 4.5 Simplified version of the western Pacific boundary in which the only land masses are New Guinea/Australia and the surrounding shelf and Indonesia/Borneo/Asia and its surrounding shelf. The ENSO pressure signal p_S originates in the western equatorial Pacific and is nearly spatially constant along and to the west of the coasts of New Guinea and northwest Australia. The pressure p_N, west of Java, is of equatorial Indian Ocean origin and differs from p_S. By geostrophy, the pressure difference along AB leads to a zonal jet flow between the two oceans. Since p_S tends to be larger than p_N, the interannual flow is correlated with ENSO. (From Clarke and Liu 1994).

south of 9°S in the Indian Ocean implies, by geostrophy, that the zonal jet is eastward and there is an anomalous flow of water into the Pacific. By similar arguments, the flow is from the Pacific into the Indian Ocean during a cold ENSO episode.

So far we have not considered the influence of interannual variability generated in the Indian Ocean on inter-ocean flow. Sea level observations and theory (Clarke and Liu 1994) show that interannual Indian Ocean zonal equatorial winds generate equatorial Kelvin waves which travel eastward to the eastern boundary of the Indian ocean. As argued earlier, the interannual sea level signal should be spatially constant along the boundary, and observations show that the equatorial interannual signal can be seen as far away as the Indian subcontinent. Thus while the zonal current at the southern tip of Java/Borneo/Asia still occurs, the strength of the inter-ocean flow is proportional to the difference between the Indian and Pacific Ocean signals. However, the Pacific Ocean ENSO signal is larger so the flow is correlated with ENSO in the way discussed earlier. Expendable bathythermograph observations between Australia and Indonesia (Meyers 1996) have confirmed the main features of the dynamics described. The upper ocean interannual transport (peak-to-peak 5 Sverdrups) is correlated with the ENSO signal and has an amplitude comparable to that predicted theoretically.

4.3. WAVE REFLECTION AT THE EASTERN PACIFIC OCEAN BOUNDARY

Figure 4.6 suggests that the ENSO climate signal in the eastern equatorial Pacific propagates poleward along the western coastline of North America. Changes in ocean climate associated with this ENSO signal have a dominant influence on the coastal ecosystem off California (see Chelton et al. 1982 and Section 12.4). The ENSO signal can also be seen along the South American coast (Enfield and Allen 1980) and there it also has a dominant influence on coastal ecosystems (Barber and Chavez 1983). The poleward spread of the ENSO equatorial signal along the eastern Pacific Ocean boundary for many thousands of kilometers is part of the equatorial reflection process at the eastern ocean boundary. In this section we will develop theory for eastern ocean boundary reflection at low frequencies. While a formal infinite sum solution can be found for any frequency for a north–south boundary (Section 4.3.1), to understand the physics of the reflection at low frequencies it is better to analyze an approximate low-frequency near-boundary solution for both north–south (Sections 4.3.2–4.3.6) and

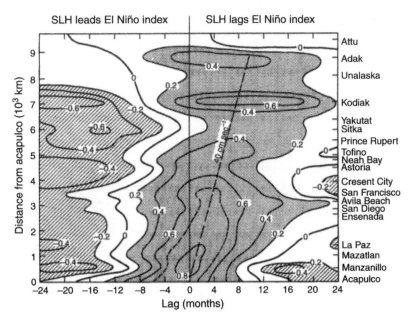

Fig. 4.6 Contour plot of the correlation between low-frequency coastal sea level in month ($t+$ lag) at each of 20 tide gauge stations and low-frequency eastern tropical Pacific sea surface temperature in month t. The monthly time series are low frequency in that they have been filtered twice with a 13-month running mean. The shaded region corresponds to correlation values >0.2; hatched regions correspond to correlation values <-0.2. The dashed line represents approximately $40\,\mathrm{cm\,s^{-1}}$ northward propagation. The 95% significance level corresponds to correlation of 0.35. (From Chelton and Davis 1982).

non-meridional (Section 4.3.7) boundaries. Also, the structure of the near-coast flow cannot be understood without consideration of bottom friction and continental shelf and slope bottom topography (Section 4.3.8). We begin the analysis with the formal infinite sum solution.

4.3.1. Moore's solution for the reflection of an equatorial Kelvin wave from a meridional eastern ocean boundary

Suppose an equatorial Kelvin wave with frequency $1 - (1/\sqrt{2}) < \omega < 1 + (1/\sqrt{2})$ propagates eastward and hits an eastern ocean boundary. According to (3.72) with $m = 1$ and the equatorial wave dispersion diagram (Fig. 3.6), in this frequency range no westward group velocity equatorial waves exist, i.e., no waves are reflected from the boundary. How is this possible? Mathematically, for this range of ω, $[\omega^2 + \omega^{-2}/4 - (2m+1)]^{1/2}$ is always purely imaginary and k is complex in (3.66); thus the equatorial wave solutions (see (3.62)), which are proportional to $\exp(ikx)$, can decay westward from the boundary. The incident equatorial Kelvin wave energy is thus trapped to the boundary and, we shall see, forms poleward propagating coastal Kelvin waves. For fixed $\omega < 1 - (1/\sqrt{2})$ a finite number of equatorial Rossby waves reflect and the rest of the energy goes into the poleward propagating coastal Kelvin waves.

Following Moore (1968), consider the reflection of an equatorial Kelvin wave of non-dimensional frequency ω from a meridional eastern ocean boundary at $x = 0$. As for the other equatorial waves, non-dimensionalize horizontal velocity by c, horizontal length by $\sqrt{c/\beta}$, p by c^2 and time by $1/\sqrt{\beta c}$. Then the non-dimensional zonal velocity field for the incoming equatorial Kelvin wave is of the form

$$u_{inc} = a \exp\left[i\omega(x - t)\right] \psi_0(y). \tag{4.13}$$

The u field is a sum of the possible low-frequency reflected waves (see (3.75)):

$$u_{Ref} = \sum_{m=1}^{\infty} \frac{1}{2} i a_m \exp[ik_m x - i\omega t]$$
$$\times \left[2^{1/2}(m+1)^{1/2}\psi_{m+1}/(\omega - k_m) + (2m)^{1/2}\psi_{m-1}(\omega + k_m)^{-1}\right] \tag{4.14}$$

where k_m from (3.66) is chosen so that the group velocity is westward or the wave remains finite. Therefore

$$k_m = -\frac{1}{2\omega} + \left[\omega^2 + \frac{1}{4\omega^2} - (2m+1)\right]^{1/2}$$

for

$$\omega^2 + \frac{1}{4\omega^2} > (2m+1) \tag{4.15a}$$

and

$$k_m = -\frac{1}{2\omega} - i\sqrt{(2m+1) - \left(\omega^2 + \frac{1}{4\omega^2}\right)}$$

for

$$\omega^2 + \frac{1}{4\omega^2} < (2m+1). \tag{4.15b}$$

As for the meridional western boundary, the condition $u_{\text{inc}} + u_{\text{Ref}} = 0$ at $x = 0$ enables us to find the unknown coefficients a_m from the orthogonality property of the eigenfunctions. From this condition and (4.13) and (4.14) we get

$$a_1 = 2^{1/2} \mathrm{i}(\omega + k_1)a \tag{4.16a}$$

$$a_{2m+1} = -a_{2m-1}\left[(\omega + k_{2m+1})/(\omega - k_{2m-1})\right]$$
$$\times [2m/(2m+1)]^{1/2} \, m = 1, 2, \ldots \tag{4.16b}$$

$$a_{2m} = 0 \quad m = 1, 2, \ldots \tag{4.16c}$$

and so the solution has been obtained. The even-numbered waves do not contribute to the reflection (see (4.16c)) because their u field has an odd symmetry in y and so cannot match the equatorial Kelvin wave u field which has an even symmetry in y. Notice that the lower the frequency, the more integers m satisfy (4.15a), i.e., the more Rossby waves are available for reflection.

Although the reflection solution has been obtained, it is not clear what the solution looks like. In order to determine and understand its structure, we will adopt another approach, determining the solution near the boundary at low frequencies (Section 4.3.3). But first we provide some background for this solution by considering the coastal Kelvin wave.

4.3.2. The coastal Kelvin wave

Consider flow near a meridional boundary on an f-plane, i.e., the Coriolis parameter f is constant and equal to βy_0. The non-dimensional equations corresponding to the f-plane problem are (cf. (3.54)–(3.56))

$$u_t - y_0 v = -p_x \tag{4.17a}$$

$$v_t + y_0 u = -p_y \tag{4.17b}$$

$$p_t + u_x + v_y = 0. \tag{4.17c}$$

Analogous to the equatorial Kelvin wave, the coastal Kelvin wave has one velocity component $\equiv 0$. With $u \equiv 0$, the coastal boundary condition is satisfied and equations (4.17) become

$$v = \frac{p_x}{\gamma_0} \tag{4.18a}$$

$$v_t = -p_y \tag{4.18b}$$

$$p_t + v_y = 0. \tag{4.18c}$$

From the last two equations

$$p_{tt} - p_{yy} = 0 \tag{4.19}$$

and this has a wave solution

$$p = A(x) \exp(\pm i\omega y - i\omega t). \tag{4.20}$$

Equation (4.18c) then implies

$$p = \pm v \tag{4.21}$$

and so by (4.18a) one has

$$\pm v_x - \gamma_0 v = 0. \tag{4.22}$$

Consequently, the solution is

$$v = \text{const.} \exp\left(\pm i\omega y - i\omega t \pm \gamma_0 x\right). \tag{4.23}$$

Equation (4.23) represents a wave (the coastal Kelvin wave) propagating along the coast with speed unity, being trapped near the coast with trapping scale $1/\gamma_0$. In dimensional form (4.23) is

$$v = \text{const.} \exp\left(i\omega y/c - i\omega t + fx/c\right) \tag{4.24}$$

with $f = \beta \gamma_0$ and c positive in the northern hemisphere and negative in the southern hemisphere so that c/f is always positive and the solution therefore decays with distance from the boundary. This choice for c implies that the coastal Kelvin wave propagates poleward along the eastern ocean boundary.

Why, physically, should the wave propagate poleward? Just like the equatorial Kelvin wave, the coastal Kelvin wave is a gravity wave because

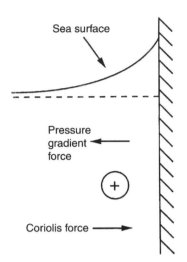

Fig. 4.7 A vertical, seaward cross-section perpendicular to the coastal wall of a meridional eastern boundary of an ocean in the northern hemisphere. The exponentially
decaying sea level corresponds to a crest of a poleward propagating coastal Kelvin
wave. Since the coastal Kelvin wave propagates northward and the current under a
gravity wave crest is in the direction of wave propagation, the current under the crest
is northward (designated by a circled plus sign in the diagram). The arrows show
that in this situation the Coriolis force and pressure gradient force can balance. If
the propagation were equatorward, then the current would be southward (out of the
paper) and the Coriolis and pressure gradient forces would be in the same direction
and would not balance.

its speed c depends on gravity and is the same as the speed of a long gravity
wave. Since gravity waves have current velocity in the direction of wave
propagation underneath a crest (see Section 3.4.1) the Coriolis force in
the northern hemisphere will be directed toward the coast underneath a
crest if the wave propagates poleward (see Fig. 4.7). In this situation the
Coriolis force can balance the pressure gradient force tending to flatten the
sea surface. However, if the wave were to propagate equatorward, both
Coriolis and pressure gradient forces would be in the same direction, no
balance would be possible and the wave could not exist. Similar arguments
can be given for the trough and southern hemisphere.

4.3.3. Solution for β-plane motions near the eastern boundary

As mentioned earlier, the formal solution of Section 4.3.1 does not
easily enable us to see its structure. Eastern boundary solutions that are
harmonic in time have been given under various restrictions by Moore

(1968), Schopf et al. (1981), Clarke (1983), Grimshaw and Allen (1988) and Clarke (1992). In Sections 4.3.3–4.3.5 we will follow Clarke (1992) and derive a solution under a low-frequency near-boundary approximation.

The non-dimensional equations corresponding to this problem are (cf. (3.54)–(3.56))

$$u_t - \gamma v = -p_x \qquad (4.25a)$$

$$v_t + \gamma u = -p_y \qquad (4.25b)$$

$$p_t + u_x + v_y = 0. \qquad (4.25c)$$

Differentiating (4.25a) with respect to t and adding to y times (4.25b) gives

$$\left(\frac{\partial^2}{\partial t^2} + \gamma^2 \right) u = -p_{xt} - \gamma p_y. \qquad (4.26a)$$

Similarly, differentiating (4.25b) with respect to t and subtracting y times (4.25a) results in

$$\left(\frac{\partial^2}{\partial t^2} + \gamma^2 \right) v = \gamma p_x - p_{yt}. \qquad (4.26b)$$

Since $\dfrac{\partial}{\partial t} = -i\omega$, the condition $u = 0$ on $x = 0$ implies from (4.26a) that at (and near) the eastern boundary $p_{xt} \sim \gamma p_y$ and hence that

$$\frac{x \text{ scale}}{y \text{ scale}} \sim \frac{\omega}{\gamma}. \qquad (4.27)$$

Thus at and near the eastern boundary and far enough away from the equator (y large enough), the x scale $\ll$ the y scale. From (4.26) and (4.27), with error $\left(\omega/\gamma \right)^2$,

$$\gamma v = p_x \qquad (4.28)$$

and

$$u = -\frac{p_y}{\gamma} - \frac{p_{xt}}{\gamma^2}. \qquad (4.29)$$

Substitution of (4.28) and (4.29) into (4.25c) leads to the field equation (with $\partial/\partial t = -i\omega$)

$$p_{xx} + \frac{i p_x}{\omega} - \gamma^2 p = 0. \qquad (4.30)$$

The solution of (4.30) is of the form

$$p = A(y)\exp(\nu(y)\cdot x)\exp(-i\omega t) \tag{4.31}$$

with

$$\nu(y) = \frac{-i}{2\omega} \pm \sqrt{y^2 - \frac{1}{4\omega^2}}. \tag{4.32}$$

For an eastern ocean boundary

$$\nu(y) = \frac{-i}{2\omega} + \sqrt{y^2 - \frac{1}{4\omega^2}} \qquad \text{for} \quad |y| \geq \frac{1}{2\omega} \tag{4.33a}$$

$$= \frac{-i}{2\omega} + i\sqrt{\frac{1}{4\omega^2} - y^2} \qquad \text{for} \quad |y| \leq \frac{1}{2\omega}. \tag{4.33b}$$

Equation (4.33a) follows because we require a finite solution away from the boundary. To verify the sign in (4.33b), first note that because the flow is due to the boundary, the energy should travel away from this source rather than head towards it. Consequently, the group velocity must be westward. From (4.32) and (4.31), with $\nu = ik$,

$$k = \frac{-1}{2\omega} \pm \sqrt{\frac{1}{4\omega^2} - y^2}$$

and $\dfrac{\partial\omega}{\partial k} < 0$ only when we choose the positive sign. Equation (4.33b) now follows.

Note that at $|y| = \dfrac{1}{2\omega}$ the solution changes from one which propagates offshore without decay to one that decays away from the boundary. The y for which $|y| = \dfrac{1}{2\omega}$ corresponds to a 'critical latitude'; the ω for which $|y| = \dfrac{1}{2\omega}$ is called a 'critical frequency.'

To find $A(y)$ in (4.31), use the boundary condition $u = 0$ at $x = 0$ which, by (4.29), is

$$p_y + p_{xt}/y = 0 \qquad \text{on} \quad x = 0. \tag{4.34}$$

Substitution of (4.31) into (4.34) gives

$$\frac{dA}{dy} - i\omega v(y)A/y = 0. \tag{4.35}$$

For $|y| \geq \dfrac{1}{2\omega}$, (4.35) is

$$\frac{dA}{dy} - [1/(2y) + i(4\omega^2 y^2 - 1)^{1/2}/2y]A = 0.$$

Define

$$\zeta = 2\omega|y|. \tag{4.36}$$

Then the equation for A, valid for $\zeta \geq 1$, can be written as

$$\frac{dA}{d\zeta} - \left[\frac{1}{2\zeta} + \frac{i}{2\zeta}\sqrt{\zeta^2 - 1}\right]A = 0.$$

Thus

$$[\ln A]_1^\zeta = \int_1^\zeta \left[\frac{1}{2\zeta} + \frac{i}{2\zeta}(\zeta^2 - 1)^{1/2}\right] d\zeta$$

and

$$A(\zeta) = 2^{-1/2}A(1)\left[(\zeta+1)^{1/2} - i(\zeta-1)^{1/2}\right]\exp\left[\left(\frac{i}{2}\right)(\zeta^2 - 1)^{1/2}\right] \quad \zeta \geq 1. \tag{4.37a}$$

This result is most easily checked by taking the logarithm of the above expression and differentiating it. Similarly

$$A(\zeta) = 2^{-1/2}A(1)\left[(1+\zeta)^{1/2} + (1-\zeta)^{1/2}\right]\exp\left[\left(-\frac{1}{2}\right)(1-\zeta^2)^{1/2}\right] \quad \zeta \leq 1. \tag{4.37b}$$

Thus the complete solution is given by (4.31) with $A(y)$ and $v(y)$ being defined in (4.37) and (4.33). The above solution only applies near the boundary and when $(\omega/y)^2 \ll 1$.

4.3.4. Limiting forms of the near-boundary low-frequency solution

In order to understand the near–boundary solution physically, we consider the form of the solution for both large and small ζ.

(i) $\zeta = 2\omega|y| \gg 1$

For large ζ our solution is

$$p \approx A(1)(2\omega|y|)^{1/2} \exp(i\omega|y| - i\omega t - i\pi/4) \exp(|y|x - ix/2\omega). \quad (4.38)$$

Notice the resemblance between this solution and that for a coastally trapped coastal Kelvin wave on an f-plane (see (4.21) and (4.23)). The essential changes are the modification of the amplitude by a factor $\sqrt{|y|}$ and the inclusion of the phase term $\exp(-ix/2\omega)$. The $\sqrt{|y|}$ factor arises because, as there is no dissipation, the energy flux of the Kelvin wave must remain constant (Moore 1968). As the coastal Kelvin wave propagates poleward, its dimensional trapping scale c/f decreases and the energy is confined to an ever narrowing region $\sim c/f$ near the coast. Since the energy flux is proportional to the amplitude of the wave squared, to keep the energy flux constant, the amplitude squared must increase like $|f|$, i.e., the amplitude must increase like $|y|^{1/2}$ for $f = \beta y$.

The term $\exp(-ix/2\omega)$ in (4.38) arises due to the conservation of potential vorticity. Using (4.28), (4.30) can be written as

$$v_{xt} + v - yp_t = 0$$

or, in dimensional form,

$$v_{xt} + \beta v - fp_t/c^2 = 0 \qquad (4.39)$$

where, just for this section of the text, v and p are dimensional. Let the north–south particle displacement be described by the function of time $\Delta y(t)$, the particle position being

$$y = \Delta y(t)$$

so that

$$v = \frac{d}{dt}(\Delta y)$$

and (4.39) can be written as

$$v_x + \beta \Delta y - fp/c^2 = 0. \tag{4.40}$$

In (4.40) v_x is the relative vorticity, $\beta \Delta y$ the planetary vorticity change when a particle is displaced Δy northward of its equilibrium position and fp/c^2 the vortex stretching. The latter follows because fw_z is fp_t/c^2 when separated into vertical modes.

Since $v = p_x/f$, when $\beta = 0$ in (4.40) we just have an equation leading to the coastal Kelvin wave solution with exponential decay. For $\beta \neq 0$, the shear in v changes to account for the planetary vorticity change $\beta \Delta y$; the necessary change in shear gives rise to the factor $\exp(-i\beta x/2\omega)$ which, non-dimensionally, is $\exp(-ix/2\omega)$.

(ii) $\zeta = 2\omega |y| \ll 1$

For small ζ our solution is

$$p \approx 2^{1/2} e^{-1/2} A(1) \exp\left(-i\omega y^2 x - i\omega t\right). \tag{4.41}$$

What does this solution represent physically? The dispersion relationship is

$$\omega/k = -1/y^2$$

or, dimensionally with $f = \beta y$,

$$\omega/k = -\beta c^2/f^2. \tag{4.42}$$

This is the dispersion relation for long, westward propagating non-dispersive Rossby waves.

Equation (4.41) shows that pressure is constant along the boundary in the limit $\zeta \to 0$. Physically, this is due to the flow being nearly geostrophic at low frequencies; there can be no pressure difference along the boundary so that there is no flow into it. Note also that, for the limiting solution (4.41) and, more generally, $\zeta < 1$, the phase is constant along the coastal boundary. In other words, the alongshore wave number ℓ is zero and the alongshore propagation speed ω/ℓ is infinite. This is in marked contrast to the large ζ case in which phase increases linearly along the coast due to coastal Kelvin wave propagation.

To summarize, the general solution (4.31) reduces to a coastal Kelvin wave propagating along the boundary at large $\zeta = 2\omega|y|$ and long, westward propagating Rossby waves for small ζ. These solutions merge at $\zeta = 1$. When $\zeta < 1$, there is no propagation at finite speed along the coastal boundary.

4.3.5. Eastern ocean near-boundary physics

The above analysis indicates that, for a given latitude, below a certain critical frequency (see (4.33))

$$\omega = 1/(2|\gamma|)$$

trapped Kelvin wave-like motion cannot exist near the boundary. Why, physically, should this be so? Consider again the vorticity balance in (4.39) and (4.40). When the frequency is high enough the terms proportional to β are negligible compared to the others and Kelvin wave dynamics is valid. As the frequency is lowered, however, the term involving β becomes relatively more important until Rossby wave dynamics takes over and the motion is no longer trapped to the boundary. An equivalent argument is to say that because

$$v = \frac{\mathrm{d}}{\mathrm{d}t}(\Delta\gamma) = -\mathrm{i}\omega\,\Delta\gamma,$$

for a given velocity amplitude, the northward particle displacement Δy increases as ω decreases. Therefore a coastal particle is displaced further along the coast as ω decreases and it experiences a greater change in planetary vorticity. Eventually, the induced change in planetary vorticity becomes dynamically important as $\beta\Delta y$ becomes comparable to the relative vorticity v_x or vortex stretching fp/c^2 in (4.40). At this point the Rossby wave mechanism described by Pedlosky (1987) takes over and energy propagates away from the boundary, i.e., the critical frequency has been reached.

What about the physics of critical latitudes – why should lower latitudes favor Rossby wave motion? The generation of relative vorticity by vortex stretching is proportional to f (see the first and last terms on the left-hand side of (4.40)), so near the equator this vorticity generation mechanism is weaker. Consequently, the planetary vorticity change $\beta\Delta y$ induced in a particle by its motion along the boundary is relatively more important near the equator, i.e., Rossby wave dynamics becomes increasingly important at lower latitudes.

4.3.6. Linking the coastal and equatorial low-frequency variability

In all of the above analysis, the boundary solution (4.31), (4.33) and (4.37) is multiplied by an arbitrary constant $A(1)$. This constant must be somehow related to the amplitude of the incoming equatorial wave which supplies the energy to the boundary disturbance. In this section we determine $A(1)$ in terms of the incoming equatorial Kelvin wave amplitude and thus link the equatorial and coastal boundary solutions.

At low frequencies the motion near the equator, where $\psi_0(y)$ is non-negligible, is described by the sum of long Rossby waves and the equatorial Kelvin wave. Therefore we may write

$$p = p_K + p_R \text{ and } u = u_K + u_R, \tag{4.43}$$

the subscripts K and R referring to equatorial Kelvin and Rossby waves, respectively. We will consider low frequencies near the equator because we will use our theory and $(\omega/y)^2$ must be small.

If the mth equatorial Rossby wave has amplitude a_m, then from (3.75) and (3.77) at $x = 0$

$$p_R + u_R = i \exp(-i\omega t) \sum_{m=1}^{\infty} a_m \frac{\sqrt{2(m+1)}\,\psi_{m+1}(y)}{\omega - k_m}.$$

Therefore, by the orthogonality relation for the Hermite functions,

$$\int_{-\infty}^{\infty} (p_R + u_R)\psi_0\, dy = 0 \tag{4.44}$$

and so, from (4.43),

$$\int_{-\infty}^{\infty} [(p - p_K) + (u - u_K)]\,\psi_0\, dy = 0. \tag{4.45}$$

But near the equator at low enough frequencies, $\zeta^2 \ll 1$ and p is given by (4.41). Also, at $x = 0$, $u = 0$ and, for the non dimensional equatorial Kelvin wave (see (3.86) and (4.13)) $p_K = u_K = a_K \psi_0(y) \exp(-i\omega t)$. Therefore, from (4.45)

$$2^{1/2} A(1) e^{-1/2} \int_{-\infty}^{\infty} \psi_0\, dy = \int_{-\infty}^{\infty} 2a_K \psi_0^2\, dy,$$

i.e.,

$$A(1) = a_K e^{1/2} \pi^{-1/4} \tag{4.46}$$

where we have used the orthogonality relation for the Hermite functions and

$$\int_{-\infty}^{\infty} \psi_0\, dy = \sqrt{2}\,\pi^{1/4}.$$

From (4.41), (3.64), $p_K = a_K \psi_0(y) \exp(-i\omega t)$ and (4.46) we thus have

$$p(\text{boundary})/p(\text{Kelvin wave, equator}) = \sqrt{2}. \qquad (4.47)$$

Equation (4.47) implies that for a given knowledge of low-frequency boundary p near the equator, it is possible to monitor low-frequency equatorial Kelvin waves. This Kelvin wave record for the Pacific has been estimated since October 1908 using coastal sea level (Clarke 1992).

4.3.7. Low-frequency reflection from non-meridional eastern ocean boundaries

We have seen that when an equatorial Kelvin wave strikes a north–south boundary, part of the incoming energy is reflected and the remainder is trapped to the boundary as poleward propagating coastal Kelvin waves. For a north–south boundary, by symmetry, half of the trapped energy must propagate northward and half southward. But realistic ocean boundaries are far from being north–south. In particular, the eastern boundary of the Pacific Ocean extends east–west over more than 70° of longitude. What is the effect of a non-meridional boundary on equatorial Kelvin wave reflection? Specifically, is more or less energy reflected? How much asymmetry is there in the poleward coastal Kelvin wave energy fluxes? What is the physics of the change in reflectivity and the coastal Kelvin wave asymmetry? These questions will be addressed using theory given in Clarke (1992) and Clarke and Shi (1991).

For simplicity, consider a straight, not necessarily meridional, boundary. Define normal and tangential coordinates (n, s) as shown in Fig. 4.8. The governing equations (3.54)–(3.56) in vector form are

$$\mathbf{u}_t + y\mathbf{k} \times \mathbf{u} = -\nabla p \qquad (4.48)$$

$$p_t + \nabla \cdot \mathbf{u} = 0 \qquad (4.49)$$

where $\mathbf{k}$ is the unit vertical vector and ∇ is the horizontal gradient operator. Take $(\partial/\partial t) - y\mathbf{k} \times$ of (4.48) to get

$$\mathbf{u}_{tt} + y^2\mathbf{u} = -\nabla p_t + y\mathbf{k} \times \nabla p$$

or, with error order $(\omega/y)^2$,

$$\mathbf{u} = -\nabla p_t/y^2 + \mathbf{k} \times \nabla p/y. \qquad (4.50)$$

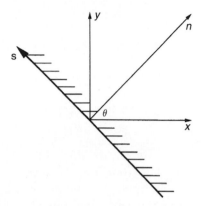

Fig. 4.8 Normal and tangential coordinates to the coastline and the angle θ.

Substitution of (4.50) into (4.49) gives

$$\nabla^2 p_t + \mathbf{i} \cdot \nabla p - 2\mathbf{j} \cdot \nabla p_t / \gamma - \gamma^2 p_t = 0 \tag{4.51}$$

where $\mathbf{i}$ and $\mathbf{j}$ are unit eastward and northward vectors, respectively. Since

$$\nabla = \mathbf{e}_n \frac{\partial}{\partial n} + \mathbf{e}_s \frac{\partial}{\partial s} \tag{4.52}$$

where $\mathbf{e}_n$ and $\mathbf{e}_s$ are unit vectors in the direction of increasing n and s, respectively, (4.51) can be written as

$$p_{nnt} + p_{sst} + \cos \theta p_n - \sin \theta p_s - 2 \sin \theta p_{nt}/\gamma - 2 \cos \theta p_{st}/\gamma - \gamma^2 p_t = 0. \tag{4.53}$$

In obtaining (4.53) we have used $\mathbf{i} \cdot \mathbf{e}_n = \cos \theta$, $\mathbf{j} \cdot \mathbf{e}_n = \sin \theta$, $\mathbf{i} \cdot \mathbf{e}_s = -\sin \theta$ and $\mathbf{j} \cdot \mathbf{e}_s = \cos \theta$, θ being the angle defined in Fig. 4.8.

At $n = 0$, $\mathbf{u} \cdot \mathbf{e}_n = 0$, i.e., from (4.50)

$$p_{nt} + \gamma p_s = 0 \tag{4.54}$$

and hence

$$n\text{-scale}/s\text{-scale} \sim \omega/\gamma. \tag{4.55}$$

Therefore, with error of order $[(\omega/\gamma)^2, (\omega \tan \theta/\gamma)]$, at or near the boundary (4.53) can be simplified to

$$p_{nnt} + \cos \theta p_n - \gamma^2 p_t = 0. \tag{4.56}$$

From (4.50)

$$v = \mathbf{u} \cdot \mathbf{e}_s = p_n/\gamma - p_{st}/\gamma^2 \approx p_n/\gamma \qquad (4.57)$$

with error $(\omega/\gamma)^2$. Hence (4.56) can be written as

$$v_{nt} + \cos\theta v - \gamma p_t = 0$$

or, in dimensional form,

$$v_{nt} + \beta \cos\theta v - f p_t/c^2 = 0 \qquad (4.58)$$

where the variables are (temporarily) dimensional and $f = \beta y$.

Let the alongshore particle displacement be described by the function of time $\Delta s(t)$, the particle position being

$$s = \Delta s(t) \qquad (4.59)$$

and so

$$v = \frac{\mathrm{d}}{\mathrm{d}t}\Delta s = -i\omega\Delta s. \qquad (4.60)$$

Substitution for v in the second term on the left-hand side of (4.58) gives

$$v_n + \beta\Delta y - f p/c^2 = 0 \qquad (4.61)$$

where the northward particle displacement Δy corresponding to the alongshore displacement Δs is

$$\Delta y = \Delta s \cos\theta. \qquad (4.62)$$

As can be seen from (4.61) and (4.62), the more the coastline is non-meridional, the smaller the planetary vorticity change $\beta\Delta y$ takes place for a given Δs and the more likely the motion is Kelvin-like. The extreme case of this is when $\theta = \pi/2$ and then the motion is always a coastally trapped Kelvin wave. Physically, in this case the coastline is east–west, particle displacements are parallel to the coast and no parcel experiences a change of planetary vorticity.

It follows from the above that the more the boundary is tilted from meridional, the more trapped motion is favored. Thus non-meridional eastern boundaries are less reflective than meridional ones. Also, we expect that the poleward propagating energy flux will not be the same for both hemispheres; the poleward energy flux will tend to be greater in the north-ern hemisphere since the Pacific eastern boundary north of the equator is less meridional and therefore favors trapped coastal Kelvin waves. More detailed calculations, done for a Pacific coastline approximated by several straight line segments, are in agreement with these results (Clarke 1992).

Solutions of (4.56) are similar to those for the meridional coastline (compare (4.56) and (4.54) with (4.30) and (4.34) with $\partial/\partial t = -i\omega$). The parameter ζ is modified to

$$\zeta = \frac{2\omega|\gamma|}{\cos\theta} \tag{4.63}$$

and, as before, when $\zeta < 1$ phase is constant along the boundary (see Clarke 1992). In dimensional form, $\zeta < 1$ corresponds to

$$\omega \leq \frac{c\cos\theta}{2|\gamma|} \tag{4.64}$$

where, temporarily, ω and $|\gamma|$ are dimensional quantities. Along most of the western coastline of the Americas, (4.64) holds at ENSO frequencies for the two lowest vertical mode phase speeds c. Therefore sea level, which is dominated by the lower vertical modes, should not propagate at finite speed along the boundary at ENSO frequencies. But observations (Enfield and Allen 1980; Chelton and Davis 1982; Fig. 4.6) show that it does. We discuss this problem in the next section.

4.3.8. Reflection of ENSO energy at the eastern Pacific Ocean boundary

The conclusion that sea level does not propagate at finite speed along the eastern Pacific Ocean boundary at ENSO frequencies is based on a model having a frictionless vertical wall coast. But real eastern boundaries are not frictionless and the shelf and slope bottom topography is not vertical. Friction should be important to the dynamics since typical frictional decay times on the shelf and slope are a few days to a few weeks, much shorter than the ENSO frequency time scale $\omega^{-1} \sim 3$ years$/2\pi = 6$ months. Clarke and Van Gorder (1994) used a coastal numerical model, forced by the ENSO frequency equatorial Kelvin wave at the equator, to examine the coastal ocean response in the presence of continental shelf and slope bottom topography and bottom friction. They found that for standard bottom friction, coastal sea level did propagate poleward at approximately the observed speed. They also found that the coastal 15 and 20°C isotherm depths propagated poleward at about $30\,\mathrm{cm\,s^{-1}}$ in agreement with the observational analysis by Kessler (1990).

Why is there poleward propagation? As already pointed out, it is not due to a coastal Kelvin wave because first and second mode coastal Kelvin waves do not exist on the Pacific Ocean eastern boundary at the very low ENSO frequencies. The physical explanation offered by Clarke and Van Gorder is illustrated in Fig. 4.9. The off-shore westward propagating ENSO frequency Rossby wave is nearly geostrophic. Consequently, the flow has associated with it a high- and low-pressure field extending perpendicular to the coast (Fig. 4.9(a)). At (say) $t = 0$ the flow is northward along the coast. In the bottom boundary layer friction weakens the current and hence the Coriolis force. Consequently the offshore pressure gradient, which does not weaken in the bottom boundary layer, forces an offshore Ekman bottom boundary layer transport. This offshore Ekman transport is balanced by a geostrophic interior flow with high coastal pressure in the south and low coastal pressure in the north. One quarter of a period later at $t = T/4$ (see Fig. 4.9(b)), the offshore propagating Rossby wave has high pressure at the coast so it appears to the coastal observer that the high pressure (high sea level) has propagated poleward along the coast from $t = 0$ to $t = T/4$. Similar

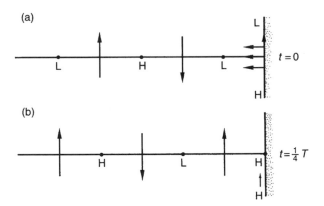

Fig. 4.9 Physics of the alongshore propagation of sea level. (a) Offshore propagating geostrophic flow at time $t=0$. The vertical arrows denote alongshore currents in geostrophic balance with the high (H) and low (L) pressure fields. The three horizontal arrows denote an offshore Ekman transport in the bottom boundary layer and the H and L at the coast the pressure field associated with the onshore geostrophic interior flow needed to balance the offshore Ekman transport. (b) Offshore propagating geostrophic flow one-quarter period later. To an observer at the coast, the high pressure (H in the south at the coast at $t=0$) appears to have propagated poleward. (From Clarke and Van Gorder 1994).

northward propagation arguments apply for low coastal pressure and sea level.

Based on their model results, Clarke and Van Gorder suggested that there should be strong ENSO currents trapped on the continental slope. However, current meter observations off Chile show that flow at ENSO frequencies is weak over the continental slope (Pizarro et al. 2001). Pizarro et al. suggest that the model error may be due to an incorrect specification of bottom friction since the ENSO current along the continental slope owes its existence to bottom friction (Clarke and Van Gorder 1994). In fact, because of buoyancy effects in the bottom frictional boundary layer, it is likely that the bottom is slippery rather than frictional (MacCready and Rhines 1991, 1993). An exception is near the coast where bottom friction remains strong because the surface and bottom mixed layers overlap and buoyancy effects in the bottom boundary layer are therefore negligible. When Pizarro et al. (2001) ran the Clarke and Van Gorder model with more appropriate bottom friction, namely, with negligible friction except near the coast, the ENSO current over the continental slope had a similar magnitude to that observed. However, such near-coast friction only causes a slight drop in interannual sea level amplitude along the coast, much smaller than that reported by Li and Clarke (2007). This bottom friction is also too small to explain the propagation speed seen in Fig. 4.6. Our understanding of interannual coastal flow is therefore still incomplete.

In our discussion of the coastal ENSO dynamics we assumed that all the ENSO ocean energy was driven remotely by ocean energy at the equator. While this is true for the South American coast (Pizarro et al. 2001) and the Central and North American coastline south of San Francisco, to the north of San Francisco coastal alongshore winds containing the ENSO signal force the coastal flow (Enfield and Allen 1980). This wind results from the large-scale extratropical atmospheric response to the ENSO heating occurring near the equator in the west-central Pacific.

The ENSO signal along the western coastline of the Americas has a major influence on the ecosystems there. This will be discussed further in Sections 12.2–12.4.

> ## REFERENCES

Barber, R. T., and F. P. Chavez, 1983: Biological consequences of El Niño. *Science*, **222**(4629), 1203–1210.

Battisti, D. S., 1988: The dynamics and thermodynamics of a warming event in a coupled tropical atmosphere–ocean model. *J. Atmos. Sci.*, **45**, 2889–2919.

Cane, M. A., and P. R. Gent, 1984: Reflection of low frequency equatorial waves at arbitrary western boundaries. *J. Mar. Res.*, **42**, 487–502.

Cane, M. A., and E. S. Sarachik, 1977: Forced baroclinic ocean motions. II. The linear equatorial bounded case. *J. Mar. Res.*, **35(2)**, 395–432.

Chelton, D. B., P. A. Bernal, and J. A. McGowan, 1982: Large-scale interannual physical and biological interaction in the California Current. *J. Mar. Res.*, **40**, 1095–1125.

Chelton, D. B., and R. E. Davis, 1982: Monthly mean sea-level variability along the west coast of North America. *J. Phys. Oceanogr.*, **12**, 757–784.

Clarke, A. J., 1982: Equatorial dynamics lecture notes. Unpublished.

Clarke, A. J., 1983: The reflection of equatorial waves from oceanic boundaries. *J. Phys. Oceanogr.*, **13**, 1193–1207.

Clarke, A. J., 1991: On the reflection and transmission of low-frequency energy at the irregular western Pacific Ocean boundary. *J. Geophys. Res.*, **96**, 3289–3305.

Clarke, A. J., 1992: Low frequency reflections from a nonmeridional eastern ocean boundary and the use of coastal sea level to monitor eastern Pacific equatorial Kelvin waves. *J. Phys. Oceanogr.*, **22**, 163–183.

Clarke, A. J., and J. Li, 2004: El Niño/La Niña shelf edge flow and Australian Western Rock Lobsters. *Geophys. Res. Lett.*, **31**, L11301, doi:10.1029/2003GL018900, 02 June 2004.

Clarke, A. J., and X. Liu, 1994: Interannual sea level in the northern and eastern Indian Ocean. *J. Phys. Oceanogr.*, **24**, 1224–1235.

Clarke, A. J., and C. Shi, 1991: Critical frequencies at ocean boundaries. *J. Geophys. Res.*, **96**, 10731–10738.

Clarke, A. J., and S. Van Gorder, 1994: On ENSO coastal currents and sea levels. *J. Phys. Oceanogr.*, **24**, 661–680.

du Penhoat, Y., and M. A. Cane, 1991: Effect of low-latitude western boundary gaps on the reflection of equatorial motions. *J. Geophys. Res.*, **96**(Suppl.), 3307–3322.

Enfield, D. B., and J. S. Allen, 1980: On the structure and dynamics of monthly mean sea level anomalies along the Pacific Coast of North and South America. *J. Phys. Oceanogr.*, **10**, 557–578.

Feng, M., G. Meyers, A. Pearce, and S. Wijffels, 2003: Annual and interannual variations of the Leeuwin Current at 32°S. *J. Geophys. Res.*, **108**(C11), 3355, doi:10.1029/2002JC 0001763, 2003.

Grimshaw, R., and J. S. Allen, 1988: Low-frequency baroclinic waves off coastal boundaries. *J. Phys. Oceanogr.*, **18**, 1124–1143.

Kessler, W. S., 1990: Observations of long Rossby waves in the northern tropical Pacific. *J. Geophys. Res.*, **95**, 5183–5217.

Kessler, W. S., 1991: Can reflected extra-equatorial Rossby waves drive ENSO? *J. Phys. Oceanogr.*, **21**, 444–452.

Li, B., and A. J. Clarke, 1994: An examination of some ENSO mechanisms using interannual sea level at the eastern and western equatorial boundaries and the zonally averaged equatorial wind. *J. Phys. Oceanogr.*, **24**, 681–690.

Li, J., and A. J. Clarke, 2007: Interannual sea level variations in the South Pacific from 5°S–28°S. *J. Phys. Oceanogr.*, **37**, 2882–2894.

MacCready, P., and P. B. Rhines, 1991: Buoyant inhibition of Ekman transport on a slope and its effect on stratified spin-up. *J. Fluid Mech.*, **223**, 631–661.

MacCready, P., and P. B. Rhines, 1993: Slippery bottom boundary layers on a slope. *J. Phys. Oceanogr.*, **23**, 5–22.

Mantua, N. J., and D. S. Battisti, 1994: Evidence for the delayed-oscillator mechanism for ENSO: The "observed" oceanic Kelvin model in the far western Pacific. *J. Phys. Oceanogr.*, **24**, 691–699.

Meyers, G., 1996: Variation of Indonesian throughflow and the El Niño–Southern Oscillation. *J. Geophys. Res.*, **101**, 12255–12263.

Moore, D. W., 1968: Planetary-gravity waves in an equatorial ocean. PhD Thesis, Harvard University, 201 pp.

Pariwono, J. I., J. A. T. Bye, and G. W. Lennon, 1986: Long-period variations of sea level in Australasia. *Geophys. J. R. Astronom. Soc.,* **87**, 43–54.

Pedlosky, J., 1987: *Geophysical Fluid Dynamics*, 2nd Edition. Springer Heidelberg, 710 pp.

Pizarro, O., A. J. Clarke, and S. Van Gorder, 2001: El Niño sea level and currents along the South American coast: Comparison of observations with theory. *J. Phys. Oceanogr.,* **31**(7), 1891–1903.

Schopf, P. S., D. L. T. Anderson, and R. Smith, 1981: Beta-dispersion of low-frequency Rossby waves. *Dyn. Atmos. Oceans,* **5**, 187–214.

Wakata, Y., and E. S. Sarachik, 1991: On the role of equatorial ocean modes in the ENSO cycle. *J. Phys. Oceanogr.,* **21**, 434–443.

CHAPTER FIVE

WIND-FORCED EQUATORIAL WAVE THEORY AND THE EQUATORIAL OCEAN RESPONSE TO ENSO WIND FORCING

Contents

5.1. OVERVIEW

In Chapters 3 and 4, we mainly considered the unforced low-frequency ocean wave motions possible near the equator. This locally unforced variability is generated by wind forcing elsewhere and here we derive the basic physics of that generation. In Sections 5.2 and 5.3, we will show that the linear, low-frequency wind-forced equatorial ocean response can be described by forced long equatorial Kelvin and Rossby waves and that a solution can be obtained using the method of characteristics. Solutions obtained for idealized wind forcing without wind-stress curl are used to understand some basic equatorial ocean physics in Section 5.4. The observed equatorial ocean sea level and thermocline height variability are then discussed in Section 5.5 and examined dynamically in Section 5.6. Section 5.7 completes the discussion of the wind-forced equatorial ocean response by describing the connection between El Niño and the equatorial Kelvin wave.

5.2. THE FORCED WAVE MODEL

Lighthill (1969) first recognized that the equatorial ocean response might be described by wind-forced long equatorial Rossby waves. Gill and Clarke (1974) formulated the equatorial ocean response in terms of the long equatorial Kelvin wave and the long equatorial Rossby waves. Cane and Sarachik (1976, 1977, 1979, 1981) considered the equatorial ocean response for various cases of wind forcing and equatorial basin geometry. We will follow the Gill and Clarke theory but use the notation of Clarke and Liu (1993).

Consider again (3.25)–(3.27), the forced equations for each vertical mode. Using a similar approach to Section 3.4, we put $f = \beta y$ and use the non-dimensionalization of Section 3.4.2. The only extra variables to consider are X_n and Y_n and both are non-dimensionalized by $c_n \sqrt{\beta c_n}$. Dropping the n subscript for notational convenience, the non-dimensional equations can be written (cf. (3.54)–(3.56))

$$u_t - yv = -p_x + X \qquad (5.1)$$

$$v_t + yu = -p_y + Y \qquad (5.2)$$

$$p_t + u_x + v_y = 0. \qquad (5.3)$$

Taking $\partial/\partial t$ of (5.1) and adding this to y times (5.2) gives

$$\left(\frac{\partial^2}{\partial t^2} + y^2\right) u = -yp_y - p_{xt} + yY + X_t. \qquad (5.4a)$$

Similarly, subtracting y times (5.1) from $\partial/\partial t$ of (5.2) gives

$$\left(\frac{\partial^2}{\partial t^2} + y^2\right) v = yp_x - p_{yt} - yX + Y_t. \qquad (5.4b)$$

We shall be interested in the *low*-frequency response of the ocean to the wind forcing and this will enable us to simplify (5.2) to the geostrophic balance (5.7) as follows. In equatorial regions, at low frequencies, the wind has a very large east–west scale (thousands of kilometers) and a much smaller north–south scale. In addition, because the surface Ekman transport is τ/f and f varies rapidly near the equator, dynamically we expect a small north–south scale response in the ocean. With these large x scale, low frequency ideas in mind, we put $\partial/\partial t \ll y$ and $\partial/\partial x \ll \partial/\partial y$ in (5.4) to get

$$y^2 u = -y p_y + y Y \tag{5.5}$$

$$y^2 v = y p_x - p_{yt} - y X. \tag{5.6}$$

In using (5.5) and (5.6) instead of (5.4), we clearly make an error at the equator at finite frequency ω because at and near the equator $\omega \gtrsim y$. But the error is small. It only occurs for very small $|y|$ (when $|y| \lesssim \omega$) with maximum error when $y = 0$. At $y = 0$, the left-hand sides of (5.5) and (5.6) are zero when there should be some tiny expression $\omega^2 u$ or $\omega^2 v$ ('tiny' in the sense that $\omega^2 u$ or $\omega^2 v \ll$ 'typical' values $y^2 u$ or $y^2 v$). Hence, the error is small.

The wind generates a non-zero pressure field, and so, in (5.6), we expect $y X \sim p_{yt}$ or $y X \sim y p_x$. Since also $Y \lesssim X$ (the north–south component of the wind stress is $\lesssim$ east–west component at the equator), in (5.5)

$$y Y / y p_y \lesssim y X / y p_y \sim p_{yt}/y p_y \quad \text{or} \quad y p_x/y p_y.$$

But $\partial p/\partial t \ll y p$ and $\partial p/\partial x \ll \partial p/\partial y$, so $y Y \ll y p_y$ in (5.5). Hence (5.5) can be reduced to

$$y u = -p_y. \tag{5.7}$$

To solve the low-frequency, large-zonal scale governing equations (5.7), (5.1) and (5.3), try a solution for v, p and u of the form

$$p = q_0 \psi_0 + \sum_{m=1}^{\infty} q_m (x, t) \left[\frac{\psi_{m+1}}{\sqrt{m+1}} + \frac{\psi_{m-1}}{\sqrt{m}} \right] \tag{5.8a}$$

$$u = q_0 \psi_0 + \sum_{m=1}^{\infty} q_m (x, t) \left[\frac{\psi_{m+1}}{\sqrt{m+1}} - \frac{\psi_{m-1}}{\sqrt{m}} \right] \tag{5.8b}$$

$$v = \sum_{m=0}^{\infty} v_m \psi_m. \tag{5.8c}$$

Why do we try a solution of this form? Firstly, we know that the $\psi_m (y)$ form a complete set. Consequently, any v field vanishing at $|y| = \infty$ can be found from an infinite sum and therefore this is a valid way to find a solution. We suspect that this may be a good way since the $\psi_m (y)$ are the 'natural' eigen functions to use as they are solutions of the unforced problem. Secondly, the y-structure for p and u, viz.,

$$\left[\frac{\psi_{m+1}}{\sqrt{m+1}} \pm \frac{\psi_{m-1}}{\sqrt{m}} \right]$$

corresponds to the y-structure of p and u in the unforced, low–frequency large–zonal scale case (see (3.80) and (3.81)).

in what follows we will make use of the Hermite function recurrence relations (3.73) and (3.78). These relations show that (5.8) satisfies (5.7) identically. It remains to find the $q_m(x, t)$ such that (5.1) and (5.3) are satisfied. Adding (5.1) and (5.3) and subtracting (5.3) from (5.1) gives

$$(p+u)_t + (p+u)_x - yv + v_y = X \qquad (5.9a)$$

$$(u-p)_t + (p-u)_x - yv - v_y = X. \qquad (5.9b)$$

Substitution of (5.8) into (5.9a) results in

$$2\psi_0(q_{0t} + q_{0x}) + \sum_{m=1}^{\infty} 2(q_{mt} + q_{mx})\psi_{m+1}/(m+1)^{1/2} - 2^{1/2}\psi_1 v_0$$

$$- \sum_{m=1}^{\infty} 2^{1/2}(m+1)^{1/2}\psi_{m+1}v_m = X \qquad (5.10)$$

where the coefficient of v_m has been found from

$$\frac{d\psi_m}{dy} - y\psi_m = -2^{1/2}(m+1)^{1/2}\psi_{m+1} \quad (m = 0, 1, 2, \ldots)$$

which follows from the subtraction of (3.73b) from (3.73a) and (3.78b) from (3.78a). Multiplying (5.10) by ψ_0 and integrating from $y = -\infty$ to $y + \infty$ gives, by the orthogonality (3.65) of the Hermite functions

$$q_{0t} + q_{0x} = \frac{1}{2}(X, \psi_0) \qquad (5.11)$$

where

$$(X, \psi_\ell) = \int_{-\infty}^{\infty} X \psi_\ell \, dy \quad \ell = 0, 1, 2, \ldots . \qquad (5.12)$$

Similarly, multiplying (5.10) by $\psi_{\ell+1}$ ($\ell = 0, 1, 2, \ldots$) and using the orthogonality of the Hermite functions gives

$$-2^{1/2} v_0 = (X, \psi_1) \tag{5.13}$$

$$2(q_{mt} + q_{mx})/(m+1)^{1/2} - (2^{1/2})(m+1)^{1/2} v_m = (X, \psi_{m+1}) \quad (m = 1, 2, \ldots). \tag{5.14}$$

Repeating a similar analysis after substitution of (5.8) into (5.9b) gives

$$2(q_{mx} - q_{mt})/m^{1/2} - v_m(m)^{1/2}\sqrt{2} = (X, \psi_{m-1}) \quad (m = 1, 2, \ldots). \tag{5.15}$$

Elimination of v_m between (5.14) and (5.15) then leads to

$$
\begin{aligned}
(2m+1)\, q_{mt} - q_{mx} = \frac{1}{2}\Big[&(\sqrt{m+1})\,(m)\,(X, \psi_{m+1}) \\
&- (\sqrt{m})\,(m+1)\,(X, \psi_{m-1}) \Big] \quad (m = 1, 2, \ldots).
\end{aligned}
\tag{5.16}
$$

Thus the solutions for p and u are given by (5.8a) and (5.8b) with $q_m(x, t)$ determined from (5.11) and (5.16).

The solution for the northward velocity v can be determined once the q_m are known. A simple form for v can be found as follows. Elimination of q_{mx} between (5.14) and (5.15) gives

$$v_m = 2\sqrt{2}\, q_{mt} - \left(X, \psi_{m+1}\sqrt{(m+1)/2}\right) + \left(X, \psi_{m-1}\sqrt{m/2}\right) \tag{5.17}$$

or, using (3.73a)

$$v_m = 2\sqrt{2}\, q_{mt} + \left(X, \frac{d\psi_m}{dy}\right) \quad (m = 1, 2, \ldots). \tag{5.18}$$

Integrating by parts then results in

$$v_m = 2\sqrt{2}\, q_{mt} + (\psi_m, -X_y) \quad (m = 1, 2, \ldots). \tag{5.19}$$

From (5.13), (3.78a) and an integration by parts, we also have

$$v_0 = (\psi_0, -X_y).$$ (5.20)

Substitution of (5.19) and (5.20) into (5.8c) then gives

$$v = 2\sqrt{2} \sum_{m=1}^{\infty} q_{mt}\psi_m + \sum_{m=0}^{\infty} (\psi_m, -X_y)\psi_m = 2\sqrt{2} \sum_{m=1}^{\infty} q_{mt}\psi_m - X_y.$$ (5.21)

The second equality in (5.21) follows by expanding $-X_y$ as $\sum_{m=0}^{\infty} a_m\psi_m$ and then using the orthogonality of the Hermite functions to get the a_m.

Notice in the above that since only X appears in the forcing terms, only the zonal wind stress τ^x is important for forcing the low-frequency flow. We will discuss this further in Section 5.4.1. Also note that when there is no forcing, (5.11) becomes

$$q_{0t} + q_{0x} = 0$$ (5.22)

with a general solution

$$q_0(x, t) = Q_0(x - t)$$ (5.23)

where Q_0 is an arbitrary differentiable function. Thus, part of the p and u solution (5.8) when there is no wind stress is

$$p = u = \psi_0 \, Q_0(x - t)$$ (5.24)

which is the non-dimensional version of a freely eastward propagating equatorial Kelvin wave (see (3.52) and (3.64)). When the zonal wind stress is non-zero, (5.22) becomes the forced wave equation (5.11), and $q_0\psi_0$ represents the forced equatorial Kelvin wave part of the response. Similar considerations for (5.16) show that the remaining part of the solution (5.8) is a sum of long Rossby waves forced by the zonal wind stress. When there is no wind forcing, these long waves travel westward at the non-dimensional speed $1/(2m+1)$ (dimensionally $c/(2m+1)$).

In summary, at large scales and at low frequencies, the equatorial ocean is driven by the zonal wind stress. The ocean response is a sum of forced long equatorial Rossby waves and a forced equatorial Kelvin wave.

5.3. SOLUTIONS FOR GENERAL LOW-FREQUENCY LARGE-SCALE WIND FORCING

The forced long-wave equations (5.11) and (5.16) can be solved by the method of characteristics. The equations are of the form

$$\frac{1}{\gamma_m}\frac{\partial q_m}{\partial t} + \frac{\partial q_m}{\partial x} = F_m(x,t) \quad (m \geq 0) \tag{5.25}$$

with the non-dimensional phase speed γ_m being unity for the equatorial Kelvin wave and $-1/(2m+1)$ for the long Rossby waves. Due to turbulent diffusion and other effects which dissipate wind generated signals in the ocean, it is advisable to include some form of damping in the model. This is most simply done by the inclusion of 'Raleigh' friction, i.e., by modifying (5.25) to

$$\frac{\lambda_m}{\gamma_m}q_m + \frac{1}{\gamma_m}\frac{\partial q_m}{\partial t} + \frac{\partial q_m}{\partial x} = F_m(x,t). \tag{5.26}$$

The damping time scale λ_m^{-1} in general depends on the mode number of the wave. Equation (5.26) can be derived in a more systematic fashion when damping is due to eddy viscosity (McCreary 1981).

The long equatorial Rossby and Kelvin waves are fundamental to the wind-forced response so it is not surprising that solutions can be readily obtained if we move along a path $x = x(t)$ such that we travel at the free wave phase speed γ_m, i.e., we move along a path with

$$\frac{dt}{dx} = \frac{1}{\gamma_m} \tag{5.27}$$

i.e.,

$$t = \frac{x}{\gamma_m} + s. \tag{5.28}$$

Then (5.26) can be written (from here on we drop the m subscript for notational convenience):

$$\frac{\lambda}{\gamma}q + \frac{dq}{dx} = F(x,t) = F\left(x, \frac{x}{\gamma} + s\right), \tag{5.29}$$

where we have used

$$\frac{dq}{dx}(x, t(x)) = \frac{\partial q}{\partial x}\frac{dx}{dx} + \frac{\partial q}{\partial t}\frac{dt}{dx} = \frac{\partial q}{\partial x} + \frac{1}{\gamma}\frac{\partial q}{\partial t}.$$

Multiplying through by the integrating factor $\exp[(\lambda/\gamma)x]$ enables us to write (5.29) as

$$\frac{d}{dx}\left[q\,\exp\left(\frac{\lambda}{\gamma}x\right)\right] = F\left(x, \frac{x}{\gamma} + s\right)\,\exp\left(\frac{\lambda}{\gamma}x\right) \qquad (5.30)$$

which, when integrated from $x = 0$ to $x = a$ gives

$$q\left(a, \frac{a}{\gamma} + s\right) = q(0, s)\,\exp\left(-\frac{\lambda}{\gamma}a\right)$$

$$+ \int_0^a F\left(x, \frac{x}{\gamma} + s\right)\,\exp\left(\frac{\lambda}{\gamma}(x - a)\right)dx. \qquad (5.31)$$

With no damping ($\lambda = 0$) and no forcing ($F \equiv 0$), (5.31) reduces to

$$q\left(a, \frac{a}{\gamma} + s\right) = q(0, s) \qquad (5.32)$$

i.e., the signal at a distance a from 0 at a time a/γ later is the same. This is consistent with a wave freely propagating at speed γ. When the damping is non-zero, the right-hand side of (5.32) is multiplied by $\exp((-\lambda/\gamma)a)$ (see (5.31) with $F \equiv 0$) so the signal is damped as it propagates. When F is non-zero between $x = 0$ and $x = a$ during the time interval $(s, s + \frac{a}{\gamma})$, the amplitude of the wave is changed by the wind stress forcing as well as the damping as it propagates. If the wind stress forcing is strong enough, free-wave propagation will not be seen.

Note that when $\gamma > 0$, we choose $a > 0$, and when $\gamma < 0$, we choose $a < 0$. In this way, $a/\gamma > 0$ and (5.31) gives, appropriately, a formula for q at a later time $t = s + a/\gamma$ given q at an earlier time $t = s$. The appropriate choice $a/\gamma > 0$ implies that the response associated with the mode having phase-speed γ is influenced only by wind stress in the direction from which the wave propagates. For example, for the equatorial Kelvin wave $\gamma = 1 > 0$ and so $a > 0$, implying that the response at $x = a$ depends only on wind stress west of $x = a$ (see (5.31)); wind stress to the east of $x = a$ does not affect the equatorial Kelvin part of the solution for an unbounded ocean. For Rossby waves, $\gamma < 0$ and so $a < 0$, implying that the response depends only on wind stress east of $x = a$.

Although wave dynamics describes the ocean response to large-scale wind forcing, often, because of the (x, y, t) structure of the forcing, freely propagating waves are not part of the response. This is the case below for some idealized wind forcing.

5.4. Solutions for Idealized Forcing

5.4.1. Unbounded ocean, spatially constant wind – the Yoshida equatorial jet

Suppose a low frequency eastward wind that is independent of x and y blows over a mid-latitude f-plane (constant f) unbounded ocean. How does the ocean respond? Since the wind is uniform and f is constant, for a wind stress τ the response will be a surface Ekman mass transport τ/f to the right of the wind for a Northern Hemisphere ocean and to the left of the wind for a Southern Hemisphere ocean (see, e.g., Gill 1982, Section 9.2). The Ekman flow is horizontally non divergent and so no flow is induced beneath the Ekman layer.

But what happens if the same spatially constant eastward wind blows over an unbounded equatorial ocean? What does the response look like in this case? Since f^{-1} is large in magnitude near the equator and changes sign at the equator, the equatorward Ekman transport is large near the equator. The convergence should cause increased sea level, lowered thermocline and a geostrophically balanced eastward equatorial flow. Although the ocean is very sensitive to zonal equatorial winds, no convergence occurs for a uniform meridional wind. This explains why only the zonal wind stress is important for large-scale low-frequency equatorial flow.

Explicit solutions can be obtained using (5.8), (5.11) and (5.16). Since the solution is the same for all x, put $\partial/\partial x = 0$ in (5.11) and (5.16). We obtain, for an initial state of rest,

$$q_0(t) = \frac{1}{2} \int_0^t (X, \psi_0)\, dt = \frac{1}{2} \left(\int_0^t X\, dt \right) I_0 \tag{5.33}$$

and

$$(2m+1)\, q_m(t) = \frac{1}{2} \left(\int_0^t X\, dt \right) \left[m\,(m+1)^{1/2}\, I_{m+1} - (m+1)\, m^{1/2}\, I_{m-1} \right] \tag{5.34}$$

where

$$I_j = \int_{-\infty}^{\infty} \psi_j \, dy. \tag{5.35}$$

The integral I_j is zero for j odd while for j even

$$I_0 = \sqrt{2} \, \pi^{1/4} \tag{5.36}$$

and

$$I_j = \left(\frac{j-1}{j} \frac{j-3}{j-2} \cdots \frac{1}{2} \right)^{1/2} \sqrt{2} \, \pi^{1/4}. \tag{5.37}$$

The solutions for p, u and v can now be found using (5.8a), (5.8b) and (5.21) with the q_m given above. They can also be determined more directly numerically by putting $\partial/\partial x = 0$ in (5.1) and (5.3) and then using these equations with (5.7) to derive a second-order ordinary differential equation for v.

The functions $u_* (y) = u / \int_0^t X \, dt$, $p_* (y) = p / \int_0^t X \, dt$, and $v_* (y) = v / X$ are shown in Fig. 5.1. As expected, even though the zonal wind stress is uniform in y, because of the strong convergence of Ekman mass transport τ^x / f at the equator, there is more mass, higher upper ocean pressure and a geostrophically balanced 'Yoshida jet' (Yoshida 1959) there. Since the surface divergence is weak away from the equator, p and u become small for large $|y|$ and X in the zonal momentum equation is balanced by $-yv$, i.e., the solution far from the equator approaches the Ekman transport non-horizontally divergent solution we discussed earlier for f constant. The north–south velocity v is asymmetric because the Ekman transports are in opposite directions in each hemisphere. No wave propagation can be seen because the ocean is forced everywhere.

5.4.2. The steady ocean response when the wind stress has no curl

Since ENSO variability is dominated by low (interannual) frequencies, it is of interest to examine the dynamics when $\partial/\partial t = 0$. We can obtain the steady solution from the q_m but it is easier to solve the governing equations directly. Since we will compare the theoretical results with those for the equatorial Pacific (see Section 5.5), and since the baroclinic mode of a two-layer ocean model is qualitatively appropriate there, we will solve the dimensional versions of (5.1), (5.2) and (5.3) for this mode. It follows

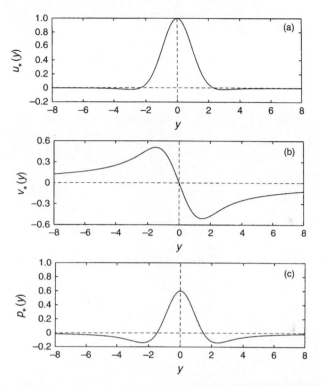

Fig. 5.1 The north south structure of (a) the Yoshida jet, $u_*(y) = u/\int_0^t X\,dt$; (b) the associated v structure $v_*(y) = v/X$ and (c) the associated p structure $p_*(y) = p/\int_0^t X\,dt$. The function $u_*(y)$ is unity at the equator because at the equator $u_t = X$ (see (5.1) with $y = 0$ and $p_x = 0$).

from (3.25)–(3.27), (3.28a) and (3.44) that the baroclinic mode equations, after dropping the 'one' subscript, are

$$u_t - fv = g\varepsilon h_x + \tau^x/(\rho_* H_1) \tag{5.38}$$

$$v_t + fu = g\varepsilon h_y + \tau^y/(\rho_* H_1) \tag{5.39}$$

$$-h_t H/(H_1 H_2) + u_x + v_y = 0. \tag{5.40}$$

Under the same large-zonal scale, low-frequency approximations used to obtain (5.7), (5.39) reduces to

$$fu = g\varepsilon h_y. \tag{5.41}$$

Our governing equations for quasi-steady flow are thus (5.41) and (5.38) and (5.40) with $\partial/\partial t = 0$:

$$-fv = g\varepsilon h_x + \tau^x/(\rho_* H_1) \tag{5.42}$$

$$u_x + v_y = 0. \tag{5.43}$$

Upon subtracting $\partial/\partial y$ of (5.42) from $\partial/\partial x$ of (5.41) and using (5.43) we obtain

$$v = -(\tau^x)_y/(\rho_* H_1 \beta) \tag{5.44}$$

which reduces to

$$v \equiv 0 \tag{5.45}$$

since the wind stress curl is zero. Therefore, from (5.43) we also have

$$u_x \equiv 0. \tag{5.46}$$

But we saw in Chapter 4 that $u = 0$ at the eastern ocean boundary is a good approximation at low frequencies so, using (5.46) we deduce

$$u \equiv 0. \tag{5.47}$$

It also follows from (5.45) and (5.42) that

$$g\varepsilon h_x = -\tau^x/(\rho_* H_1). \tag{5.48}$$

This may also be written in terms of sea-level gradient using (3.45) and the approximation that $H_2 \approx H$ for the equatorial Pacific:

$$\rho_* g \eta_x = \tau^x/H_1. \tag{5.49}$$

Physically, (5.49) describes a balance between the eastward wind stress anomaly that tilts sea level up toward the east and the opposing westward pressure gradient associated with the tilted sea level. When the eastward wind stress anomaly pushes the surface layer of water eastward, the thermocline tilts downward according to (5.48). This downward tilt corresponds to an anomalous increase in the thermocline depth by an amount $D = -h$. Hence (5.48) may also be written

$$\rho_* g\varepsilon \frac{\partial D}{\partial x} = \tau^x/H_1. \tag{5.50}$$

In summary, when there is no wind stress curl and the forcing frequency is low enough so that the ocean response is quasi-steady, the velocity is zero ((5.45) and (5.47)) and the zonal pressure gradient balances the wind stress forcing ((5.48), (5.49) and (5.50)).

5.5. OBSERVED LOW-FREQUENCY BEHAVIOR OF THE EQUATORIAL PACIFIC SEA LEVEL AND THERMOCLINE

Interannual equatorial Pacific Ocean variability plays a key role in ENSO dynamics. In this section, we will discuss Pacific sea level and thermocline depth anomalies averaged over the region 5°S–5°N. The thermocline depth will be approximated by the depth of the 20°C isotherm.

Figures 2.2 and 2.9 suggest that the ENSO wind that is effective in forcing the ocean is largely confined to the west/central equatorial Pacific between about 160°E and 150°W. More recent satellite data suggest (see Fig. 5.2) that between 5°S and 5°N, at least for January 1992–December 2000, the zonal wind-stress anomalies may have a slightly greater zonal extent (from about 155°E to 140°W). Equatorial Kelvin and low meridional mode Rossby waves propagate across this distance in a time short compared to interannual variability, suggesting that the ocean response might be quasi-steady. Based on this and (5.50), we would expect the anomalous depth of the equatorial 20°C isotherm to tilt over the approximate longitude range 155°E–140°W and to be horizontal elsewhere. We can check this using empirical orthogonal function (EOF) analysis (see Appendix A) of $D_{Eq}(x, t)$, the anomalous 20°C isotherm depth averaged over the equatorial strip 5°S–5°N. Such an analysis indicates that most

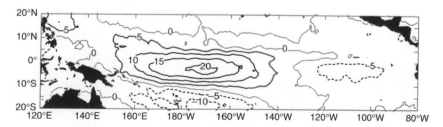

Fig. 5.2 Regression of the European Remote-Sensing (ERS) zonal wind stress anomaly onto the El Niño index NINO3.4. Negative contours are dashed, positive contours are solid, and the zero line is a solid line thinner than those for the positive contours. The contours are in intervals of 5 mPa/°C. (From Clarke et al. 2007).

of the variance is described by the first two EOF modes (51% for mode
1 + 35% for mode 2 = a total of 86%). Hence

$$D_{Eq}(x, t) \approx S_1(x)\phi_1(t) + S_2(x)\phi_2(t). \tag{5.51}$$

Figure 5.3 shows that $\phi_1(t)$ is in phase with NINO3.4 and hence the zonal
equatorial wind stress anomalies (see Fig. 5.2), consistent with a quasi-
steady response of the equatorial ocean to low-frequency wind forcing. As
expected from (5.50), $S_1(x)$ tilts upward in the central Pacific from about
160°E to 120°W, in approximately the same region as the westerly wind

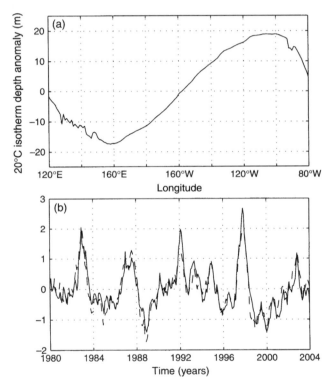

Fig. 5.3 (a) First EOF (51% of the variance) of the 20°C isotherm depth anomaly
(m) for the equatorial (5°S–5°N) Pacific and (b) the corresponding first principal
component (solid line). The principal component time series has been normalized to
have variance = 0.5 so that (a) represents the amplitude of the isotherm depth anomaly.
The dashed curve in (b) is the NINO3.4 time series normalized so that it has the same
variance as the first principal component. The correlation coefficient between the two
time series is r = 0.92. The isotherm data were supplied by Neville Smith (1995a,b).
(From Clarke et al. 2007).

anomalies in Fig. 5.2 (about 155°E–140°W). However, near the boundary where the zonal wind stress in Fig. 5.2 is small and, based on (5.50), we expect the thermocline slope to be small, it is not (see Fig. 5.3a east of 100°W and west of 160°E). Near the boundaries, particularly near the eastern equatorial Pacific boundary, the 20°C isotherm depth is not in a fixed proportional relationship with the sea level because their EOF structure functions vary differently there (compare Figs 5.3 and 5.4). This indicates that the simple two–density layer baroclinic model based on the 20°C isotherm depth is invalid near the eastern boundary. Possibly this is due to the heat flux anomalies documented by Holland and Mitchum (2003); these will change the 20°C isotherm depth thermodynamically, ruining its ability to represent a dynamical constant density surface. In addition, the two–density layer model here may not be a good approximation to the real

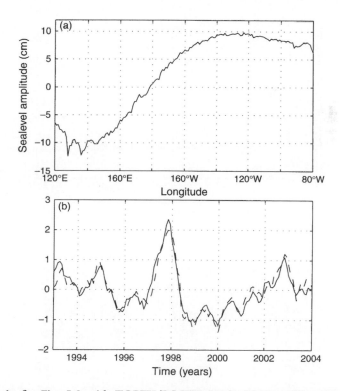

Fig. 5.4 As for Fig. 5.3 with TOPEX/POSEIDON/JASON satellite-estimated sea level anomalies (cm) replacing the 20°C isotherm depth anomalies. In the sea-level case, the first EOF explains 69% of the variance, and the correlation of the first principal component with NINO3.4 is r = 0.95. (From Clarke et al. 2007).

continuous stratification anyway because in the eastern equatorial Pacific, the finite thickness thermocline intersects the surface.

Although the 20°C isotherm depth anomaly does not represent the wind tilt dynamics (5.50) near the boundary, equivalent tilt dynamics (5.49) is approximately valid for sea level even near the eastern equatorial boundary. Specifically, the sea-level amplitude not only tilts appropriately upward in the region 155°E–140°W but also, in the eastern equatorial Pacific, there is a slight downward slope (see Fig. 5.4a), consistent with the weak easterly anomalies present during El Niño (see Fig. 5.2).

The second EOF $S_2(x)\,\phi_2(t)$ in (5.51) explains 35% of the variance (see Fig. 5.5), comparable to the 51% explained by the first EOF. This second main mode of equatorial ocean ENSO response was first found observationally by Meinen and McPhaden (2000) in their EOF analysis of 20°C isotherm depth anomalies for the tropical Pacific. By contrast with the first EOF, whose structure function $S_1(x)$ is of opposite sign in the eastern

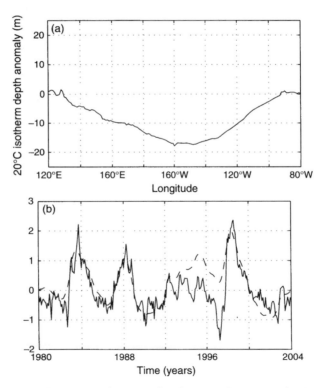

Fig. 5.5 (a,b) As for Fig. 5.3 but now for the second EOF, explaining 35% of the variance. In this case, the dashed curve is not NINO3.4 but rather $\int_0^t \text{NINO3.4}\,(t_*)dt_*$. The integral time series has zero mean and variance 0.5. (From Clarke et al. 2007).

and western Pacific, the second EOF has a structure function $S_2(x)$ essentially of one sign right across the Pacific (compare Fig. 5.3a with Fig. 5.5a). The principal components also differ in that $\phi_1(t)$ is approximately proportional to NINO3.4 while $\phi_2(t)$ is such that $d\phi_2/dt \approx$ NINO3.4 (see Figs 5.3b and 5.5b). In fact, by the properties of EOFs, $S_1(x)$ and $S_2(x)$ are spatially uncorrelated and $\phi_1(t)$ and $\phi_2(t)$ are uncorrelated in time; clearly a different physical mechanism is responsible for the coherent, large-scale second EOF contributions to the equatorial ocean response. We shall see in the next section that this part of the response is due to the wind-stress curl.

5.6. ENSO EQUATORIAL OCEAN RESPONSE DUE TO THE WIND STRESS CURL

The second part of the response, $S_2(x)\phi_2(t)$ in (5.51), cannot be described by quasi-steady dynamics as was the case for $S_1(x)\phi_1(t)$ because $\phi_2(t)$ is not in phase with NINO3.4 and the equatorial wind (Fig. 5.5b). When the quasi-steady approximation is not made, the governing equations are (5.38), (5.41) and (5.40). Since $H_2 \approx H$ and $D = -h$, these equations may be written

$$u_t - fv = -g\varepsilon \frac{\partial D}{\partial x} + \frac{\tau^x}{(\rho_* H_1)} \tag{5.52}$$

$$fu = -g\varepsilon \frac{\partial D}{\partial y} \tag{5.53}$$

$$D_t + H_1(u_x + v_y) = 0. \tag{5.54}$$

To gain insight into the second part $S_2(x)\phi_2(t)$ of the response, we begin by integrating (5.54) over the 5°S–5°N equatorial strip S where large-scale dynamics is valid. Specifically, we integrate from the western equatorial boundary east of the narrow western boundary layer flows and the Indonesian throughflow to the eastern ocean boundary. Using the divergence theorem, we obtain

$$\int_\Gamma \mathbf{e}_n \cdot \mathbf{u} \, ds + \int_{5°N} v \, dx - \int_{5°S} v \, dx + \int_S D_t \, dS/H_1 = 0 \tag{5.55}$$

where $\mathbf{e}_n$ is the unit outward normal to the curve Γ from 5°S–5°N just eastward of the western boundary flows. Calculations (Clarke et al. 2007) suggest that (5.55) is invalid if we use the 20°C isotherm depth to represent D. This is because an anomalous heat flux can change the isotherm depth and so the isotherm depth does not represent a material surface. But calculations by Clarke et al. 2007 also show that the anomalous warm water volume contributed by the heat flux is approximately balanced by the first term on the left-hand side of (5.55). In what follows, for simplicity, we will keep D as the 20°C isotherm depth and omit the first term on the left-hand side of (5.55).

Since the strip S is east of the western boundary flows, we can estimate v at 5°S and 5°N in (5.55) from the governing large-scale equations (5.52)–(5.54). Differentiating (5.53) with respect to x and subtracting (5.52) differentiated with respect to y gives

$$-u_{yt} + \beta v + f(u_x + v_y) = \text{curl } \tau/(\rho_* H_1) \tag{5.56}$$

where curl $\tau = (-\tau^x)_y$. It follows from (5.53) and (5.54) that (5.56) can also be written

$$v = \text{curl } \tau/(\rho_* H_1 \beta) + f D_t/(H_1 \beta) - \left(g\varepsilon \frac{\partial D}{\partial y} \middle/ f\right)_{yt} /\beta. \tag{5.57}$$

The ratio of the third term on the right-hand side of (5.57) to the second is of order

$$(g\varepsilon H_1 D_y/f)_y/f \, D \sim c^2/(f^2 L_y^2) \tag{5.58}$$

where c is the baroclinic mode phase speed (see (3.40) with $H_2 \approx H$) and the y scale L_y is set by the distance of 5°N and 5°S from the equator. Using $c = 2.7 \, \text{ms}^{-1}$ in (5.58) gives $c^2/(f^2 L_y^2) \approx 0.15 \ll 1$. The analysis of Kessler and McPhaden (1995) confirms that the third term on the right-hand side of (5.57) at 5°N and 5°S can be omitted without serious error. Therefore, when (5.57) is substituted into (5.55) we have, approximately, that

$$\frac{\partial}{\partial t}\left(\int_S D \, dS\right) + \frac{\partial}{\partial t}\left(\frac{f_N}{\beta}\int_{5°N} D \, dx - \frac{f_S}{\beta}\int_{5°S} D \, dx\right)$$
$$= -\frac{1}{(\rho_* \beta)}\left[\int_{5°N} \text{curl } \tau \, dx - \int_{5°S} \text{curl } \tau \, dx\right] \tag{5.59}$$

where f_N and f_S refer to the values of the Coriolis parameter at 5°N and 5°S. If Δy is the meridional distance of 5°N from the equator, then $f_N = \beta \Delta y = -f_S$ and so the second term inside the brackets on the left-hand side of (5.59) is

$$\frac{f_N}{\beta}\int_{5°N} D \, dx - \frac{f_S}{\beta}\int_{5°S} D \, dx = \Delta y\left(\int_{5°N} D \, dx + \int_{5°S} D \, dx\right). \tag{5.60}$$

To the extent that the trapezoidal rule of integration is a good approximation of $\int_{-\Delta y}^{\Delta y} (\int D\, dx)\, dy = \int_S D\, dS$, the right-hand side of (5.60) is $\int_S D\, dS$. Calculations (Clarke et al. 2007) show that the left-hand side of (5.60) and $\int_S D\, dS$ are well correlated in time but that the left-hand side of (5.60) is equal to $\alpha \int_S D\, dS$ with $\alpha \approx 0.53$. Based on this result and (5.59), we have

$$-(1+\alpha)\frac{\partial}{\partial t}\int_S D\, dS = \frac{1}{\rho_* \beta}\left[\int_{5°N} \text{curl } \tau\, dx - \int_{5°S} \text{curl } \tau\, dx\right]. \quad (5.61)$$

Equation (5.61) describes an ocean response that is quite different from the response (5.50); in (5.50) the tilted pressure response is in phase with the wind stress but (5.61) suggests a mode which is a quarter period out of phase with the wind stress curl (and also the wind stress and NINO3.4).

Equation (5.61) can be interpreted physically using (5.56) with the small term $-u_{yt}$ omitted. During El Niño, the anomalous wind stress curl is positive on average along 5°N and negative on average along 5°S (see Fig. 5.6). Thus, during El Niño, the wind stress curl exerts a positive torque on the ocean along 5°N and a negative torque along 5°S. On the large scales of interest, the ocean responds in two ways. At 5°N it increases its angular momentum by moving northward where the planetary vorticity is larger (the βv term in (5.56)) and by expanding horizontally (f times the horizontal divergence in (5.56)). The expansion increases the angular momentum because each parcel of fluid has mass further from its local center of rotation. Similarly, at 5°S the ocean makes its angular momentum more negative by moving southward (where f is more negative) and by again expanding horizontally (since f is negative). By the continuity of mass, the expansion results in decreasing warm water at the northern and

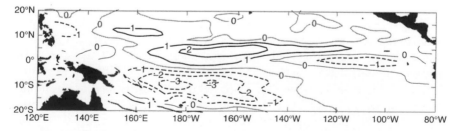

Fig. 5.6 Regression of European Remote-Sensing (ERS) wind stress curl anomaly onto the El Niño index NINO3.4. Negative contours are dashed, positive contours are solid, and the zero line is a solid line thinner than those for the positive contours. The contours are in intervals of $(10^{-8}\, \text{Pa/m})/°\text{C}$. (From Clarke et al. 2007).

southern edges of the box. Mathematically in (5.61), a positive curl anomaly at 5°N and negative curl anomaly at 5°S lead to warm water volume loss by this expansion mechanism at a rate $-\alpha \frac{\partial}{\partial t} \int_S D \, dS$. The rest of the warm water volume loss, $-\frac{\partial}{\partial t}(\int_S D \, dS)$, is due to the other part of the ocean response to the curl, namely, the meridional flow across 5°N and 5°S. During La Niña, similar physics applies but with opposite sign.

The response (5.61) is directly related to the second EOF in (5.51). To see this, begin by integrating (5.51) with respect to x from the furthest western point $(x = 0)$ of the western Pacific boundary to the furthest eastern point of the eastern Pacific boundary $(x = L)$ and then differentiating with respect to time. This results in

$$\frac{\partial}{\partial t}\left(\int_0^L D_{Eq}(x, t)dx\right) = \left(\int_0^L S_1(x)dx\right)\frac{d\phi_1(t)}{\partial t}$$

$$+\left(\int_0^L S_2(x)dx\right)\frac{d\phi_2(t)}{\partial t}. \qquad (5.62)$$

Since $S_2(x)$ is almost entirely of one sign while $S_1(x)$ changes sign (see Figs 5.3a and 5.5a), $\int_0^L S_2(x)\,dx$ is much larger than $\int_0^L S_1(x)\,dx$ in (5.62). Thus the right-hand side of (5.62) is dominated by the second EOF. Also, from the definition of $D_{Eq}(x, t)$ as D averaged meridionally over the ocean from 5°S to 5°N, we have

$$2\Delta y \int_0^L D_{Eq}(x, t) \approx \int_S D \, dS. \qquad (5.63)$$

Consequently, (5.62) reduces to

$$\frac{\partial}{\partial t}\left(\int_S D \, dS\right) \approx 2\Delta y \left(\int_0^L S_2(x)dx\right)\frac{d\phi_2}{dt}. \qquad (5.64)$$

Equations (5.64) and (5.61) show that the left-hand side of (5.64) is proportional to minus the wind stress curl forcing anomaly. Since that wind anomaly is approximately proportional to NINO3.4, we see from (5.61) and (5.64) that the left-hand side of (5.64) is proportional to minus NINO3.4. Figure 5.5a shows that $\int_0^L S_2(x)\,dx$ is negative so $d\phi_2/dt$ should be in phase with NINO3.4. This is consistent with Fig. 5.5b which shows that $\phi_2(t)$ is in phase with $\int_0^t \text{NINO3.4}(t_*)dt_*$.

In summary, the equatorial thermocline response to ENSO wind forcing is partly in the form of a thermocline tilt that is essentially in phase with the zonal wind stress (see (5.50)) and partly anomalous upper ocean warm water volume in quadrature with the wind stress curl. The equatorial warm water volume (or, equivalently, upper ocean heat content) was originally discussed by Cane and Zebiak (1985) (see also Zebiak and Cane 1987 and Jin 1997). The equatorial warm water volume appears to play an important role in coupled ENSO dynamics (see Chapter 7).

 ## 5.7. EL NIÑO AND EQUATORIAL KELVIN WAVES

In the previous section we assumed that equatorial ocean adjustment to ENSO wind forcing was rapid enough that phase lags due to equatorial wave propagation could be ignored. Here we discuss El Niño and its connection with equatorial wave propagation. Specifically, we will consider the ocean response near to the eastern equatorial Pacific Ocean boundary.

As mentioned earlier in Section 5.5, the main region of ENSO wind forcing occurs in the west/central equatorial Pacific between about 155°E and 140°W. By long wave theory, the only wave that can carry energy east of this region of forcing is an equatorial Kelvin wave since only this wave has eastward group velocity. When an El Niño occurs, the zonal wind forcing anomalies are eastward, resulting in a forced response with an anomalous eastward current since f is zero at the equator. The directly forced equatorial Kelvin wave carrying energy eastward of the forcing region therefore is associated with an eastward current anomaly and, since $u = p/c$ (see (3.52)), a positive pressure anomaly, i.e., a raised sea level and depressed thermocline.

When the equatorial Kelvin wave reaches the South American coast, it lowers the thermocline there. The remotely forced lowered thermocline near the South American coast causes the surface waters to be warmer and poorer in nutrients and El Niño results (see Figs 2.16 and 2.17). This occurs even though the winds off Peru and Ecuador remain favorable for upwelling (Wyrtki 1975). Both Godfrey (1975) and Wyrtki (1975) recognized that El Niño could be generated by remote anomalous westerly equatorial wind forcing and equatorial Kelvin wave propagation.

Integration of the forced Kelvin wave equation (5.11) from the western equatorial Pacific boundary $x = 0$ to the eastern boundary $x = L$ gives, using (5.31) with $\gamma = 1$, $\lambda = 0$ and a replaced by L,

$$q_0(L, L+s) = q_0(0, s) + \int_0^L F_0(x, x+s)\, \mathrm{d}x. \qquad (5.65)$$

This (non-dimensional) equation states that the equatorial Kelvin wave signal at the eastern equatorial Pacific boundary has two contributors.

One, the directly forced lowered thermocline contribution discussed earlier, would be the total contribution if there were no western boundary. The second contribution, $q_0(0, s)$, is due to the freely propagating equatorial Kelvin wave resulting from Rossby wave energy reflected from the western boundary. Rossby wave energy generated by the wind forcing will propagate westward from that forcing and strike the western boundary. By the arguments given earlier, the Rossby waves overall will have anomalous eastward equatorial currents which the reflected equatorial Kelvin wave will try to cancel in order that the long wave western boundary condition (4.9) is satisfied. This means that the Kelvin wave will have negative u and p, i.e., an anomalously low sea level and raised thermocline. The time taken for an equatorial Kelvin wave to cross the basin for the dominant first two vertical modes is short compared to ENSO periodicity. Therefore, when the raised thermocline Kelvin wave reflected at the western boundary reaches the eastern boundary, it is still nearly out of phase with the directly forced lowered thermocline wave and tends to cancel it. Calculations based on long records of sea level and zonal equatorial winds (Li and Clarke 1994) show that the amplitude of the El Niño at the eastern boundary disagrees with the theoretical estimate if the reflected equatorial Kelvin wave contribution from the western boundary is ignored but agrees if the reflected equatorial Kelvin wave contribution is included.

In summary, El Niño is generated remotely by interannual equatorial zonal wind stress usually between about 155°E and 140°W. The remotely forced depressed thermocline and raised sea level in the eastern equatorial Pacific is due to the sum of a directly forced downwelling equatorial Kelvin wave and a comparable but smaller upwelling one resulting from Rossby wave energy reflected at the western equatorial Pacific boundary.

5.8. CONCLUSION

In this chapter, we have seen how the ocean response to large-scale wind forcing can be described in terms of long ocean waves. We have also seen how both the wind stress and the wind stress curl influence the equatorial ocean and give rise to two very different types of thermocline response. In Chapter 7, we shall see how wind-driven currents and the thermocline displacement can influence the SST, thereby changing the heating of the atmosphere. Anomalous atmospheric heating drives anomalous winds which then change the SST, etc., so atmosphere and ocean are coupled together. However, before we can consider coupled ocean–atmosphere dynamics, we must first examine how the atmosphere responds to changes in SST (Chapter 6).

▶ REFERENCES

Cane, M. A., and E. S. Sarachik, 1976: Forced baroclinic ocean motions. I. The linear equatorial unbounded case. *J. Mar. Res.,* **34**(4), 629–665.

Cane, M. A., and E. S. Sarachik, 1977: Forced baroclinic ocean motions. II. The linear equatorial bounded case. *J. Mar. Res.,* **35**(2), 395–432.

Cane, M. A., and E. S. Sarachik, 1979: Forced baroclinic ocean motions. III. The linear equatorial basin case. *J. Mar. Res.,* **37**, 355–398.

Cane, M. A., and E. S. Sarachik, 1981: The response of a linear baroclinic equatorial ocean to periodic forcing. *J. Mar. Res.,* **39**, 651–693.

Cane, M. A., and S. E. Zebiak, 1985: A theory for El Niño and Southern Oscillation. *Science,* **228**, 1084–1087.

Clarke, A. J., and X. Liu, 1993: Observations and dynamics of semiannual and annual sea levels near the eastern equatorial Indian Ocean boundary. *J. Phys. Oceanogr.,* **23**, 386–399.

Clarke, A. J., S. Van Gorder, and G. Colantuono, 2007: Wind stress curl and ENSO discharge/recharge in the equatorial Pacific. *J. Phys. Oceanogr.,* **37**, 1077–1091.

Gill, A. E., 1982: *Atmosphere–Ocean Dynamics.* Academic Press, New York, 662 pp.

Gill, A. E., and A. J. Clarke, 1974: Wind-induced upwelling, coastal currents and sea-level changes. *Deep-Sea Res.,* **21**, 325–345.

Godfrey, J. S., 1975: On ocean spin-down. I. A linear experiment. *J. Phys. Oceanogr.,* **5**, 399–409.

Holland, C. L., and G. T. Mitchum, 2003: Interannual volume variability in the tropical Pacific. *J. Geophys. Res.,* **108**(C11), 3369, doi:10.1029/2003JC001835.

Jin, F. F., 1997: An equatorial ocean recharge paradigm for ENSO. Part I: Conceptual model. *J. Atmos. Sci.,* **54**, 811–829.

Kessler, W. S., and M. J. McPhaden, 1995: Oceanic equatorial waves and the 1991–1993 El Niño. *J. Climate,* **8**, 1757–1774.

Li, B., and A. J. Clarke, 1994: An examination of some ENSO mechanisms using interannual sea level at the eastern and western equatorial boundaries and the zonally averaged equatorial wind. *J. Phys. Oceanogr.,* **24,** 681–690.

Lighthill, M. J., 1969: Dynamic response of the Indian Ocean to the onset of the Southwest Monsoon. *Philos. Trans. R. Soc.* London, Ser. A **265**, 45–93.

McCreary, J. P., 1981: A linear stratified ocean model of the equatorial undercurrent. *Philos. Trans. R. Soc.* London, **298**, 603–635.

Meinen, C. S., and M. J. McPhaden, 2000: Observations of warm water volume changes in the equatorial Pacific and their relationship to El Niño and La Niña. *J. Climate,* **13**, 3551–3559.

Smith, N. R., 1995a: An improved system for tropical ocean subsurface temperature analysis. *J. Atmos. Oceanic Technol.,* **12**, 850–870.

Smith, N. R., 1995b: The BMRC ocean thermal analysis system. *Aust. Met. Mag.,* **44**, 93–110.

Wyrtki, K., 1975: El Niño – The dynamic response of the equatorial Pacific Ocean to atmospheric forcing. *J. Phys. Oceanogr.,* **5**, 572–584.

Yoshida, K., 1959: A theory of the Cromwell Current (the equatorial undercurrent) and of the equatorial upwelling—An interpretation in a similarity to a coastal circulation. *J. Oceanogr. Soc. Jpn.* **15**, 159–170.

Zebiak, S. E., and M. A. Cane, 1987: A model El Niño–Southern Oscillation. *Mon. Weather Rev.,* **115**, 2262–2278.

CHAPTER SIX

SEA SURFACE TEMPERATURE, DEEP ATMOSPHERIC CONVECTION AND ENSO SURFACE WINDS

Contents

6.1. OVERVIEW

In Chapters 3–5 we considered how the equatorial ocean responds to forcing by the low–frequency wind. This is the 'atmosphere drives ocean' part of the coupled ENSO problem. To understand ENSO, we must also examine how the ocean drives the atmosphere. Specifically, in this chapter we will consider how the SST anomalies are linked to ENSO precipitation and how the associated latent heating drives the surface equatorial wind anomalies which then force the ocean.

Following a brief introduction to the structure of the atmosphere (Section 6.2), we discuss the equation of state and First Law of Thermodynamics for a moist atmosphere. This leads to the use of the conservation of moist static energy to suggest why deep atmospheric convection and associated precipitation typically does not occur over water colder than about 28°C (Sections 6.3–6.5). The remainder of the chapter treats the second topic, the surface equatorial wind response to ENSO heating. Here we will concentrate on the zonal equatorial surface winds since the ocean is most sensitive to these (see Chapter 5). We begin by discussing the close relationship between heating and vertical velocity, defining the geopotential and deriving the continuity equation in pressure coordinates (Sections 6.6 and 6.7). We are then able to formulate and discuss the physics of a model for surface equatorial wind anomalies driven by ENSO deep convective heating (Section 6.8).

6.2. SOME PRELIMINARIES

The troposphere, the layer of the atmosphere closest to the earth, is about 17 km thick in the tropics. The tropical troposphere contains 80% of the mass of the tropical atmosphere and nearly all of the water vapor and clouds. In the troposphere, temperature decreases with height.

Immediately above the troposphere is the stratosphere in which temperature increases with height. This increasing temperature is due to ozone. Oxygen atoms form above 25 km in the stratosphere and combine with molecular oxygen (O_2) to form ozone (O_3). Ozone in the stratosphere strongly absorbs solar ultraviolet radiation and radiates heat, thus causing the temperature of the stratosphere to increase with height and making it very stable. The absorption of ultraviolet radiation by ozone protects the biosphere from biologically harmful ultraviolet radiation. The stratosphere extends to about 50 km above the earth's surface and the division between it and the troposphere is called the tropopause.

Our analysis will focus on the troposphere, for this is the layer of the atmosphere relevant to ENSO dynamics.

6.3. EQUATION OF STATE

6.3.1. 'Ideal' gas

To within an excellent approximation, atmospheric gases obey the ideal gas equation

$$pV = nR_*T \qquad (6.1)$$

where p is the pressure of the gas, V its volume, T its temperature, n the number of molecules of the gas in the volume V and R_* the universal gas constant. The number n of molecules is usually measured in units of kilomoles, one kilomole being 6.022×10^{26}, the number of molecules in M kilograms of a gas with molecular weight M. By this definition, if m is the mass (in kg) of the gas in the volume V then in (6.1)

$$n = m/M. \qquad (6.2)$$

6.3.2. Dry air

Air consists of four main gases: nitrogen, oxygen, argon and water vapor. Water vapor varies considerably in the atmosphere but the other constituents are well approximated by the fixed volume ratios 78.1% nitrogen, 21.0% oxygen and 0.9% argon. Since each constituent itself behaves as an ideal gas, it follows from (6.1) and (6.2) that in a volume V of dry air at temperature T

$$p_i V = m_i R_* T/M_i \qquad (6.3)$$

where p_i, m_i and M_i refer to the pressure, mass and molecular weight of the ith constituent. The total pressure p is the sum of the pressures p_i so summing (6.3) gives

$$pV = \left(\sum_{i=1}^{3} m_i/M_i \right) R_*T = m_d R_* T/M_{air} \qquad (6.4)$$

where m_d is the sum of the masses m_i of dry air and M_{air}, the apparent molecular weight of dry air, is given by

$$M_{air} = m_d \left/ \left(\sum_{i=1}^{3} m_i/M_i \right) \right. = 28.966. \qquad (6.5)$$

Equation (6.4) can also be written as

$$p = \rho R T \tag{6.6}$$

where ρ is the density of the dry air and R, the gas constant for dry air, is

$$R = R_*/M_{\text{air}} = 287.04 \text{ J kg}^{-1}\text{K}^{-1}, \tag{6.7}$$

a value found using (6.5) and universal gas constant $R_* = 8314.36 \text{ J kmol}^{-1}\text{K}^{-1}$.

6.3.3. Moist air

Consider now the more general case when we allow for moisture in the air. By analogy with (6.6), the equation of state for the water vapor of pressure p_V and density ρ_V at temperature T is

$$p_V = \rho_V R_V T. \tag{6.8}$$

In (6.8) the gas constant for water vapor

$$R_V = R_*/M_w = 461.50 \text{ J kg}^{-1}\text{K}^{-1} \tag{6.9}$$

since M_w, the molecular weight of water, is 18.016.

The density ρ of the moist air of volume V having mass m_w of water vapor and m_d of dry air is

$$\rho = \frac{m_w + m_d}{V} = \rho_V + \rho_d. \tag{6.10}$$

Defining the specific humidity q as the mass of water vapor per unit mass of moist air we see that

$$q = m_w/(m_w + m_d) = \rho_V/\rho. \tag{6.11}$$

At temperature T the partial pressure p_d of the dry air of density ρ_d satisfies

$$p_d = \rho_d R T. \tag{6.12}$$

Since the total pressure p is the sum of p_V and p_d, adding (6.8) and (6.12) gives

$$p = (\rho_d R + \rho_V R_V)T = \rho R T(\rho_d/\rho + \rho_V R_V/\rho R). \tag{6.13}$$

From (6.10) and (6.11)

$$\rho_d/\rho = 1 - q \qquad (6.14)$$

while (6.7) and (6.9) imply that

$$R/R_V = M_w/M_{air} = \varepsilon = 0.62197. \qquad (6.15a)$$

Thus (6.13) can be written, using (6.14) and (6.15a), as

$$p = \rho R T(1 - q + q/\varepsilon). \qquad (6.15b)$$

Sometimes a virtual temperature T_V is defined by

$$T_V = T(1 - q + q/\varepsilon) = T(1 + 0.6078q). \qquad (6.16)$$

Surface values of q in the tropics are $\sim 18 \times 10^{-3}$ so T_V can differ from T by about 1% or about 3 K. For most situations we may put $q = 0$ and $T_V = T$ without significant error. However, in the discussion of energy and latent heating in the next two sections, specific humidity should and will be included.

6.4. THE FIRST LAW OF THERMODYNAMICS AND MOIST STATIC ENERGY

The First Law of Thermodynamics is a special case of the Law of Conservation of Energy. It states that δE, the increase in internal energy per unit mass of a volume of fluid, is equal to the heat added per unit mass (δQ) minus the work done by the fluid per unit mass, i.e.,

$$\delta E = \delta Q - p\,\delta\alpha. \qquad (6.17)$$

In (6.17) $\alpha = 1/\rho$ is the volume per unit mass and so $p\,\delta\alpha$ is the work done by the fluid per unit mass as it expands by an amount $\delta\alpha$.

The internal energy per unit mass E is due to the kinetic energy associated with individual molecules. About 99% of dry air consists of the diatomic molecules oxygen and nitrogen. For a diatomic ideal gas, and hence approximately for dry air,

$$E = (5/2)RT. \qquad (6.18)$$

The specific heat of dry air at constant volume is defined to be

$$c_V = \frac{dE}{dT}. \tag{6.19}$$

Such a definition makes sense because when volume is constant, $\delta\alpha = 0$ in (6.17) and $\delta E = \delta Q$. In that case

$$c_V = \frac{\delta E}{\delta T} = \frac{\delta Q}{\delta T} \tag{6.20a}$$

and for $\delta T = 1°K$, $c_V = \delta Q$, i.e., the specific heat at constant volume is the amount of heat needed to raise the temperature of a unit mass by $1°K$. It follows from (6.18) and (6.19) that

$$c_V = (5/2)\ R = 717.6\ \text{J kg}^{-1}\text{K}^{-1}. \tag{6.20b}$$

When the more realistic case of moist air is considered, we must take into account that water is triatomic instead of diatomic as for almost all dry air molecules. For triatomic molecules the factor 5/2 in (6.18) changes to 3 so

$$E = 3R_V T. \tag{6.21}$$

Since moist air is a mixture of dry air and water vapor in the proportions $1 - q$ and q per unit mass of moist air, we have that the internal energy per unit mass of moist air is, from (6.18) and (6.21),

$$E = (5/2)\ RT(1 - q) + 3R_V Tq. \tag{6.22}$$

Using (6.15a) we see that (6.22) may also be written as

$$E = (5/2)\ RT(1 - q + 6q/(5\varepsilon)). \tag{6.23}$$

To see how the First Law of Thermodynamics is linked to moist static energy, we must combine (6.17) with the equation of state for moist air. Using $\alpha = 1/\rho$, (6.15b) and (6.16) this equation is

$$p\alpha = RT_V. \tag{6.24}$$

For small changes

$$p\,\delta\alpha + \alpha\,\delta p = \delta(RT_V). \tag{6.25}$$

Eliminating $p\delta\alpha$ between (6.25) and (6.17) gives

$$\delta(E + RT_V) - \alpha\,\delta p = \delta Q. \tag{6.26}$$

At constant pressure for dry air the above equation reduces, using (6.23), to

$$\delta((7/2)RT) = \delta Q, \tag{6.27}$$

i.e.,

$$(7/2)R = \left(\frac{\delta Q}{\delta T}\right)_{p\ \text{constant}}. \tag{6.28}$$

Just as $c_V = (\delta Q/\delta T)$ at constant volume, so for dry air, by (6.28), the specific heat at constant pressure

$$c_p = \left(\frac{\delta Q}{\delta T}\right)_{p\ \text{constant}} = (7/2)\ R = 1004.6\ \text{J kg}^{-1}\,\text{K}^{-1}. \tag{6.29}$$

Using (6.16), (6.23) and (6.29), we see that the expression $(E + RT_V)$ in (6.26) can be written as

$$E + RT_V = c_p T(1 - q + 8q/(7\varepsilon)) = c_p T(1 + 0.8375q). \tag{6.30}$$

Substitute (6.30) into (6.26), divide by the time interval δt and let δt approach zero following a parcel of moist air. We obtain

$$\frac{D}{Dt}(c_p T\,(1 + 0.8375q)) - \alpha\frac{Dp}{Dt} = \dot{Q} \tag{6.31}$$

where $\dot{Q}$ is the rate of heating per unit mass. This is one form of the thermodynamic energy equation. We will use this equation later when we describe the atmospheric response to ENSO heating (see Sections 6.6, 6.8 and Chapter 9).

Under some approximations, the thermodynamic energy equation (6.31) leads to the conservation of moist static energy. First, for the large horizontal scale low-frequency motions of interest to us, hydrostatic balance

$$p_z = -g\rho \tag{6.32}$$

is an excellent approximation. This balance is valid both for pressure fluctuations due to the motion and when the air is at rest. The vertical gradient of the total air pressure thus includes a contribution from the pressure when there is no flow. Because pressure fluctuations due to the motion are much smaller than the pressure of the air at rest, wp_z is the dominant term in Dp/Dt. Therefore, we may write the Dp/Dt term in (6.31) as

$$\frac{Dp}{Dt} \approx wp_z = -w\rho g = -wg/\alpha \qquad (6.33)$$

where $\alpha = \rho^{-1}$. Second, when $\dot{Q}$ is due to latent heating,

$$\dot{Q} = -L\frac{Dq}{Dt} \simeq \frac{D}{Dt}(-Lq) \qquad (6.34)$$

where L is the latent heat of condensation $\approx 2.44 \times 10^6$ J kg^{-1}. Substitution of (6.33) and (6.34) into (6.31) and using $D(gz)/Dt = gw$ gives

$$D/Dt(c_p T(1 + 0.8375q) + Lq + gz) = 0. \qquad (6.35)$$

The conserved quantity $m_{SE} = c_p T (1 + 0.8375q) + Lq + gz$ is known as the moist static energy, static because kinetic energy is negligible and has not been included. In the next section we will use the conservation of moist static energy to help explain why the minimum sea surface temperature needed for deep atmospheric convection is about 28°C.

6.5. SEA SURFACE TEMPERATURE AND DEEP ATMOSPHERIC CONVECTION

In the tropics, especially near the surface, the air is conditionally unstable, i.e., without water vapor it would be stable but it is potentially unstable when its water vapor is taken into account. Specifically, if an air parcel is displaced upward it will expand, do work and cool. The cooling may cause the water vapor to condense and, if it does, the latent heat may warm the parcel enough so that it continues to rise, i.e., the air is unstable. The instability is conditional; it depends on the air being warm and moist enough and the air parcel being displaced high enough for condensation to occur. Since the temperature and moisture in air parcels is determined by the underlying SST, one expects that a *necessary* condition for deep atmospheric convection to occur is that the SST must be high enough. Observational analyses by Krueger and Gray (1969), Gadgil et al. (1984), Graham and Barnett (1987), Waliser et al. (1993) and Zhang (1993) found that usually SST must be greater than about 27.5–28°C before tropical large-scale deep atmospheric convection can occur.

We can check this idea using the conservation of moist static energy m_{SE} and the relationship between SST, specific humidity and air temperature. At the 200 hPa level, observations (Sud et al. 1999) show that typically

$$m_{SE}/c_p \approx 347.5\,\text{K}. \tag{6.36}$$

If air parcels near the surface are to reach the 200 hPa level, then at say a height of 10 m,

$$m_{SE}/c_p \geq 347.5\,\text{K},$$

i.e., with $gz = 98\,\text{m}^2\,\text{s}^{-2}$, we have from (6.35) that

$$T(1 + 0.8375q) + Lq/c_p + 98/c_p \geq 347.5\,\text{K}. \tag{6.37}$$

To proceed further we must specify the 10 m height air temperature T and the specific humidity q in terms of the SST. In the western Pacific Godfrey et al. (1991) found that at 10 m the air temperature was about 1.35 K less than the sea surface temperature so that we may write

$$T \approx 273.15 + T_{SS} - 1.35 \tag{6.38}$$

where T_{SS} is the sea surface temperature in °C. Wang (1988) determined that the near-surface mixing ratio $q/(1-q) \approx q$ is highly correlated with the SST and that, for SST in °C,

$$q/(1-q) \approx q = ((0.940)\,T_{SS} - 7.64) \times 10^{-3}. \tag{6.39}$$

Substitution of (6.38) and (6.39) into (6.37) gives, with $L = 2.44 \times 10^6\,\text{J}\,\text{kg}^{-1}$ and $c_p = 1004.6\,\text{J}\,\text{kg}^{-1}$,

$$0.940(T_{SS})^2 + 4168\,T_{SS} - 114,504 \geq 0$$

which is valid when

$$T_{SS} \geq 27.31°\text{C} \approx 27.3°\text{C}. \tag{6.40}$$

This temperature may be a little low because some moist static energy is lost through entrainment of drier colder air as parcels of air rise. Sud et al. estimate that therefore the moist static energy required should be increased by about $1800\,\mathrm{J\,kg^{-1}}$, i.e., the right-hand side of (6.37) should be increased by $1800/c_p = 1.79\,\mathrm{K}$. This causes (6.40) to be modified to

$$T_{SS} \geq 27.81°\mathrm{C} \approx 27.8°\mathrm{C}. \qquad (6.41)$$

Equation (6.41) is consistent with the observed approximate temperature necessary for large-scale deep atmospheric convection.

This minimum temperature of about 28°C necessary for deep atmospheric convection has direct application to ENSO dynamics. Figure 2.2(b) shows that the biggest El Niño SST anomalies are in the eastern equatorial Pacific. However, the deep convective heating is not observed there (see Fig. 2.2(c)) because the water is usually colder than 28°C. Consequently the deep atmospheric convection anomalies are further to the west near the edge of the western equatorial Pacific warm pool where water temperatures are 28°C and above (see Fig. 2.4).

As pointed out earlier, a minimum SST of about 28°C is only a necessary rather than sufficient condition for deep atmospheric convection. However, at the edge of the equatorial warm pool SST exceeding about 28°C seems sufficient for deep convection. This can be seen in Fig. 6.1 which shows that deep atmospheric convection (as indexed by the $240\,\mathrm{W\,m^{-2}}$ equatorial outgoing long wave radiation isoline) moves zonally back and forth with the warm water (as indexed by the equatorial 28.5°C isotherm).

The air/sea interaction occurring at the edge of the western equatorial Pacific warm pool is the key region of ocean–atmosphere coupling occurring in ENSO. It is here that the atmosphere most strongly drives the ocean through zonal equatorial surface wind anomalies and it is here that the atmosphere is being most strongly driven by anomalous deep atmospheric convective heating resulting from the strong linkage to warm water SST anomalies. This chapter has been concerned with the 'ocean drives atmosphere' part of the coupled problem and so far we have considered the linkage between SST and deep atmospheric convection; the next task is to understand why such anomalous heating causes the westerly surface wind anomalies seen in Fig. 2.2. But before we can examine a model to understand the surface wind response, we need to consider the relationship between heating and vertical velocity (Section 6.6) as well as geopotential, pressure coordinates and the continuity equation (Section 6.7).

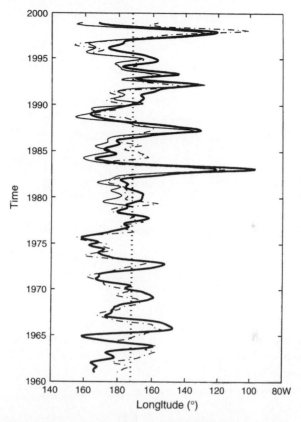

Fig. 6.1 Zonal displacement of the equatorial 28.5°C SST isotherm (heavy solid line: a proxy of the east edge of the warm pool), zonal displacement of the equatorial $-4\,\mathrm{m\,s}^{-1}$ westerly surface isotach (dash–dot line) and zonal displacement of the $240\,\mathrm{W\,m}^{-2}$ equatorial OLR isoline (thin solid line). The 'equatorial' SST and the equatorial surface wind are actually an average from 4°N to 4°S and the equatorial OLR an average from 5°N to 5°S. The vertical dotted line is the mean position of the 28.5°C isotherm (172.2°W). Monthly values of SST, wind and OLR have been smoothed with a double 5-month running mean filter. The correlation coefficients are 0.94 for SST and OLR isolines, 0.76 for OLR and wind isolines and 0.73 for wind and SST isolines. (After Shu and Clarke 2002.)

6.6. Heating and Vertical Velocity

It follows from (6.31), (6.33) and c_p constant that

$$c_p \frac{\mathrm{D}}{\mathrm{D}t}\left(T\left(1+0.8375q\right)\right)+wg = \dot{Q},$$

i.e.,

$$(1 + 0.8375q)\frac{DT}{Dt} + \frac{wg}{c_p} = \frac{\dot{Q}}{c_p} - 0.8375T\frac{Dq}{Dt}. \tag{6.42}$$

Since $0.8375q$ is always about 2% or less, the first term on the left-hand side of (6.42) can be well approximated by DT/Dt. The last term on the right-hand side of (6.42) is also negligible because

$$\left| 0.8375T\frac{Dq}{Dt} \right| << \left| \frac{\dot{Q}}{c_p} \right| \sim \left| L\frac{Dq}{Dt} \middle/ c_p \right|$$

since $0.8375T << L/c_p \approx 2400\,\mathrm{K}$. Thus (6.42) simplifies to

$$\frac{DT}{Dt} + \frac{wg}{c_p} = \frac{\dot{Q}}{c_p}. \tag{6.43}$$

Because temperature fluctuations due to the flow are much smaller than the mean changes of temperature with height in a state of rest, for the large horizontal scales of interest we have $DT/Dt \approx w\,\partial T/\partial z$. Consequently (6.43) reduces to

$$w\left(\frac{\partial T}{\partial z} + \frac{g}{c_p} \right) \approx \frac{\dot{Q}}{c_p}. \tag{6.44}$$

As $g/c_p \approx 9.8\,\mathrm{K\,km^{-1}}$ while $\partial T/\partial z \approx -6.5\,\mathrm{K\ km^{-1}}$, (6.44) shows us that w is positive when the atmosphere is heated and that parcels of air that lose heat will sink. Equation (6.44) can be understood physically if, based on (6.43), we rewrite (6.44) as

$$\frac{\partial T}{\partial t} \approx 0 \approx \frac{\dot{Q}}{c_p} - w\left(\frac{\partial T}{\partial z} + \frac{g}{c_p} \right).$$

Thus the cooling experienced locally due to rising air

$$\left(\frac{\partial T}{\partial t} = -w\left(\frac{\partial T}{\partial z} + \frac{g}{c_p} \right) < 0 \right)$$

is balanced by the diabatic heating

$$\left(\frac{\partial T}{\partial t} = \frac{\dot{Q}}{c_p} > 0\right)$$

so that there is negligible temperature change. The adiabatic cooling $-w\left((\partial T/\partial z) + (g/c_p)\right)$ has two contributions that partly cancel; a parcel of air displaced upward will cause local warming because temperature decreases with height $(\partial T/\partial t = -w(\partial T/\partial z) > 0)$ but this is overcome by a greater cooling as heat content $c_p T$ per unit mass is converted into potential energy gz per unit mass as the parcel rises:

$$\left(\frac{\partial \left(c_p T\right)}{\partial t} = \frac{-D(gz)}{Dt}, \quad \text{i.e.,} \quad \frac{\partial T}{\partial t} = -w\frac{g}{c_p}\right).$$

6.7. GEOPOTENTIAL, PRESSURE COORDINATES AND THE CONTINUITY EQUATION

6.7.1. Geopotential and geopotential height

In the preceding sections it was necessary to use the equation of state for moist air in order to estimate a minimum necessary temperature for deep atmospheric convection. In most applications, however, the equation of dry air is sufficiently accurate and we will use it in what follows.

Define the geopotential Φ as the work done in raising a unit mass to a height z from mean sea level:

$$\Phi = \int_0^z g \, dz. \tag{6.45}$$

Then it follows from (6.45), the hydrostatic equation

$$\frac{dp}{dz} = -g\rho \tag{6.46}$$

and the equation of state (6.6) for dry air that

$$d\Phi = g \, dz = -\frac{dp}{\rho} = -\frac{RT \, dp}{p} = -RT \, d \ln p. \tag{6.47}$$

Define the geopotential height as

$$Z = \Phi(z)/g_0 \text{ where } g_0 = 9.80665 \,\mathrm{m\,s^{-2}} \tag{6.48}$$

is the global average value of the acceleration due to gravity at mean sea level. Thus in the troposphere and lower stratosphere Z is numerically almost identical to the geometric height z. Integration of (6.47) between pressure surfaces p_1 and p_2 yields

$$\Delta Z = Z_2 - Z_1 = \frac{R}{g_0} \int_{p_2}^{p_1} T \,\mathrm{d}\,\ln p \tag{6.49}$$

where ΔZ is the thickness of the atmospheric layer between the pressure surface p_2 and p_1. Note that if we define a mean layer temperature $\overline{T}$ by

$$\overline{T} = \int_{p_2}^{p_1} T \,\mathrm{d}\ln p \bigg/ \int_{p_2}^{p_1} \mathrm{d}\ln p \tag{6.50}$$

then

$$\Delta Z = H \,\ln(p_1/p_2) \tag{6.51}$$

where the scale height is

$$H = R\overline{T}/g_0. \tag{6.52}$$

Therefore the thickness ΔZ of a layer between two pressure surfaces is proportional to the mean temperature $\overline{T}$ of that layer.

Between the surface and the 70 km level, the absolute temperature for the standard atmosphere is within 15% of a constant value of $T = 250\,\mathrm{K}$. If we take T constant and equal to 250 K in (6.49), then with $p_1 = p_0$ corresponding to the earth's surface at $Z = 0$ and with Z_2, p_2 replaced by Z, p we have

$$Z = -\frac{R\overline{T}}{g_0} \ln\left(\frac{p}{p_0}\right) = -H \,\ln\frac{p}{p_0},$$

i.e.,

$$p(Z) = p(0) \exp(-Z/H). \tag{6.53}$$

With T constant, it follows from (6.6) and (6.53) that

$$\rho(Z) = \rho(0) \,\exp(-Z/H). \tag{6.54}$$

For $\overline{T} = 250\,\text{K}$, $R = 287\,\text{J}\,\text{K}^{-1}\,\text{kg}^{-1}$ and $g_0 = 9.8\,\text{m}\,\text{s}^{-2}$, the scale height $H = 7.3\,\text{km}$. Therefore density varies markedly over the approximately 18 km depth of the tropical troposphere.

Because of this density variation it is helpful to use pressure instead of height as the vertical coordinate. The main changes in the equations of motion are in the continuity equation and that gradients of pressure at constant height become gradients of height at constant pressure.

6.7.2. Horizontal gradients of Φ

Let the height of some constant pressure surface be defined by $z = z'(x, y)$. Since the pressure does not change on this surface

$$0 = \frac{\partial p}{\partial x}(x, y, z'(x, y)) = \frac{\partial p}{\partial x}\frac{\partial x}{\partial x} + \frac{\partial p}{\partial z}\frac{\partial z'}{\partial x},$$

i.e.,

$$\frac{\partial p}{\partial x} - g\rho\frac{\partial z'}{\partial x} = 0 \tag{6.55}$$

where we have used the chain rule and hydrostatic balance. Now if $\delta z'$ is the change in height of the constant pressure surface over a distance δx then

$$\delta\Phi = g\,\delta z' \tag{6.56}$$

is the geopotential change of that pressure surface over the height $\delta z'$. Hence, by (6.55) and (6.56)

$$\frac{1}{\rho}\frac{\partial p}{\partial x} = \frac{\partial\Phi}{\partial x} \tag{6.57a}$$

where $\partial\Phi/\partial x$ is evaluated on a constant pressure surface and $\partial p/\partial x$ on a surface of constant height z. By similar analysis, we obtain

$$\frac{1}{\rho}\frac{\partial p}{\partial y} = \frac{\partial\Phi}{\partial y}. \tag{6.57b}$$

Since ρ is positive, (6.57) shows that horizontal gradients in p at constant height are the same sign as horizontal gradients in the height of a constant pressure surface. Thus a 'high' in pressure on a geopotential (level, constant height) surface appears as a positive height of a nearby constant

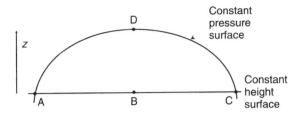

Fig. 6.2 Diagram of high pressure on a geopotential (level, constant height) surface appearing as a positive height of a nearby constant pressure surface. The constant pressure surface ADC is locally higher at D than A or C and correspondingly the pressure at B is higher than at A or C because pressure decreases with height. Local high pressure and positive geopotential height (of a pressure surface) thus go together.

pressure surface. We can see this diagrammatically in Fig. 6.2 The constant pressure surface ADC is locally higher at D than A or C and correspondingly the pressure at B is higher than at A or C because pressure decreases with height. Local high pressure and positive geopotential height (of a pressure surface) thus go together.

6.7.3. The continuity equation in pressure coordinates

We have seen that the horizontal pressure gradient term in the momentum equations can be simplified if the momentum equations are written in isobaric coordinates. The continuity equation can also be simplified. To see this, consider a control volume

$$\delta V = \delta x \, \delta y \, \delta z \tag{6.58}$$

of fixed mass δM that moves with the fluid. From hydrostatic balance

$$\delta M = \rho \, \delta x \, \delta y \, \delta z = -\delta x \, \delta y \, \delta p/g. \tag{6.59}$$

Since the mass of the fluid element is conserved following the motion

$$\frac{\mathrm{d}}{\mathrm{d}t}\delta M \bigg/ \delta M = 0$$

or, by (6.59), for g constant

$$\frac{\mathrm{d}}{\mathrm{d}t}(\delta x \, \delta y \, \delta p) \bigg/ (\delta x \, \delta y \, \delta p) = 0. \tag{6.60}$$

This can be simplified to

$$\frac{1}{\delta x}\frac{d}{dt}\delta x + \frac{1}{\delta y}\frac{d}{dt}\delta y + \frac{1}{\delta p}\frac{d}{dt}\delta p = 0.$$

Now $(d/dt)\delta x$ is the rate of change of the fluid element in the x direction with time and this is equal to the difference in velocity in the x direction between each end (in x) of the fluid element. Thus

$$\frac{d}{dt}(\delta x) = \delta u \quad \text{and} \quad \frac{d}{dt}(\delta y) = \delta v. \tag{6.61}$$

Hence the continuity equation is

$$\frac{\delta u}{\delta x} + \frac{\delta v}{\delta y} + \frac{\delta \omega}{\delta p} = 0 \tag{6.62}$$

where

$$\omega = \frac{dp}{dt} \tag{6.63}$$

is the rate of change of pressure following the motion. Letting δx, δy and δp approach zero implies that the continuity equation is

$$\frac{\partial u}{\partial x} + \frac{\partial v}{\partial y} + \frac{\partial \omega}{\partial p} = 0. \tag{6.64}$$

6.8. A MODEL OF THE NEAR-SURFACE TROPICAL ATMOSPHERE'S RESPONSE TO LARGE-SCALE CONVECTIVE HEATING

As mentioned earlier in Chapter 2, the key region for ENSO generation is in the west–central equatorial Pacific from about 150°E to 150°W. It is here, near the edge of the western equatorial Pacific warm pool, that the ocean is most strongly driven by anomalous winds and where the atmosphere is most strongly driven by anomalous heating. Crucial to the ocean–atmosphere coupling is the westerly (easterly) surface wind response to anomalous atmospheric heating (cooling), because it is this response that fuels the ENSO instability described in Chapter 7. Specifically, when the equatorial warm pool in the western equatorial Pacific (see Fig. 2.4) is displaced slightly eastward, the precipitation travels with it and so, via latent heating, the atmosphere is anomalously heated to the east of the normal position of the equatorial warm water. This anomalous heating generates anomalous westerly equatorial winds (see Fig. 2.2(c)) and these winds push the warm water farther to the east, causing a bigger eastward equatorial warm water displacement. This in turn creates more anomalous heating and further growth in the instability. Gill and Rasmusson (1983) first described this coupled ocean–atmosphere instability which is discussed in Section 7.2.2. It is crudely modeled mathematically there by (7.2) with right-hand side given by (7.3).

Note that the coupled ocean–atmosphere instability depends fundamentally on the direction of the anomalous zonal wind. For example, if anomalous easterly (instead of westerly) winds resulted from anomalous heating, then any small eastward displacement of the warm pool of water would be pushed back to the west, the small eastward perturbation would be damped out and ENSO as observed would not occur. But why does anomalous ENSO heating (cooling) drive westerly (easterly) surface winds? In this section we develop a model to understand this. The model is essentially the Neelin (1988) surface frictional boundary layer model which uses some of the ideas in Matsuno (1966) and Gill (1980).

6.8.1. Model formulation

In the turbulent atmospheric boundary layer near the ocean surface the horizontal momentum equation is

$$\frac{D\mathbf{u}}{Dt} + f\mathbf{k} \times \mathbf{u} = -\nabla p/\rho + \mathbf{X}_z/\rho \tag{6.65}$$

where now $\mathbf{u}$ is the horizontal wind velocity and $\mathbf{X}$ the stress caused by friction at the surface. We will assume that, for the long time, large spatial scales of interest, $D\mathbf{u}/Dt$ is negligible in the surface frictional layer. Then (6.65) can be written in isobaric coordinates as

$$f\mathbf{k} \times \mathbf{u} = -\nabla\Phi - g\,\mathbf{X}_p \tag{6.66}$$

where we have used (6.57) and

$$\mathbf{X}_z = \mathbf{X}_p\,p_z = -g\rho\mathbf{X}_p. \tag{6.67}$$

Integrate (6.66) with respect to p from the top of the frictional boundary layer (where $\mathbf{X} = 0$ and $p = p_\mathrm{T}$) to the surface (where $\mathbf{X} = \mathbf{X}_\mathrm{bot}$ and $p = p_\mathrm{B}$). We get

$$f\mathbf{k} \times \mathbf{U} = -\nabla\phi - \mathbf{X}_\mathrm{bot} \qquad (6.68)$$

where

$$\mathbf{U} = \int_{p_\mathrm{T}}^{p_\mathrm{B}} \mathbf{u}\,\frac{\mathrm{d}p}{g} = \int_{h}^{z=0} \mathbf{u}(-\rho\,\mathrm{d}z) = \int_{z=0}^{h} \mathbf{u}\rho\,\mathrm{d}z \qquad (6.69)$$

is the boundary mass transport and

$$\phi = \int_{p_\mathrm{T}}^{p_\mathrm{B}} \Phi\,\frac{\mathrm{d}p}{g}. \qquad (6.70)$$

Write

$$\mathbf{X}_\mathrm{bot} = \rho c_\mathrm{D}\,|\mathbf{u}|\mathbf{u} = K\mathbf{U} \qquad (6.71)$$

where c_D is a drag coefficient. An estimate for K is

$$\frac{\rho c_\mathrm{D}|\mathbf{u}|^2}{|\mathbf{U}|} \approx \frac{\rho c_\mathrm{D}|\mathbf{u}|^2}{h|\mathbf{u}|\rho} = \frac{c_\mathrm{D}|\mathbf{u}|}{h}. \qquad (6.72)$$

Using $c_\mathrm{D} = 2 \times 10^{-3}$, $|\mathbf{u}| = 5\,\mathrm{m\,s^{-1}}$ and $h = 1000$ m, $K^{-1} \approx 1$ day. Thus (6.68) can be written as

$$K\mathbf{U} + f\mathbf{k} \times \mathbf{U} = -\nabla\phi. \qquad (6.73)$$

To complete the model equations, integrate the continuity equation (6.64) with respect to p to obtain

$$U_x + V_y + \left[\frac{\omega}{g}\right]_{p_\mathrm{T}}^{p_\mathrm{B}} = 0. \qquad (6.74)$$

Now because $w = 0$ at the surface which is approximated by $p = p_\mathrm{B}$, we have from (6.33) that $\omega = Dp/Dt = 0$ at $p = p_\mathrm{B}$. Hence from (6.74),

$$U_x + V_y = \omega_\mathrm{T}/g \qquad (6.75)$$

where ω_T is the value of ω at the top of the boundary layer $p = p_\mathrm{T}$.

Equations (6.33) and (6.44) show that ω is proportional and of opposite sign to the vertical velocity and heating rate Q. The latter can be written as

$$\dot{Q} = \dot{Q}_{LH} + \dot{Q}_0 \qquad (6.76)$$

where $\dot{Q}_{LH}$ is the latent heating and $\dot{Q}_0$ all other diabatic heating. Consider first the latent heating and suppose that ω due to latent heating is of the form

$$\omega(x, y, p) = F(p)\omega_H(x, y) \qquad (6.77)$$

for some function F. In regions of deep convection, latent heating occurs over most of the troposphere with a maximum at about the 400 mb level so we expect F to be a positive function with a maximum at that level. The precipitation rate $P(x, y)$ is proportional to an integral of the latent heating rate over the depth of the troposphere and therefore to an integral of ω. Under (6.77) this integral is proportional to $\omega_H(x, y)$ which is also proportional to $\omega_T = \omega(p_T)$. Thus we may write, for the latent heating contribution to ω_T,

$$\omega_T \text{ (latent heating)}/g = -aP \qquad (6.78)$$

for some appropriate positive constant a.

We will suppose that $\dot{Q}_0$ (see (6.76)), the anomalous heating that is not $\dot{Q}_{LH}$, behaves as a Newtonian cooling and so is proportional to the anomalous temperature $-T'$. Observations (Horel and Wallace 1981) show that T' is proportional to the geopotential height and therefore also its integral ϕ (see (6.70)). Thus $\dot{Q}_0$ is proportional to ϕ and we may write, using (6.76) and (6.78), that

$$\omega_T/g = -K_T\phi/c^2 - aP. \qquad (6.79)$$

In (6.79) we have written the proportionality constant in the form K_T/c^2 for reasons which will soon be apparent. The constant $(K_T)^{-1}$ is a thermal damping time and $c \approx 68\,\text{m s}^{-1}$ is the atmospheric Kelvin wave speed. The negative sign appears with K_T in (6.79) because when (say) $\dot{Q}_0$ is positive, ϕ and w are positive and ω, being opposite in sign to w, is negative.

Combining (6.79) with (6.75) gives

$$K_T\phi/c^2 + U_x + V_y = -aP. \qquad (6.80)$$

The model equations are thus (6.80) and (6.73). They are similar in form to (3.25)–(3.27), the equations describing the response of the equatorial ocean to wind forcing. The differences are that $\partial/\partial t$ replaces K and K_T and instead of motion being forced by the windstress through the X and Y terms, it is forced by the precipitation P. The similar structure suggests that the atmospheric response might be described by forced, damped, equatorial Kelvin and Rossby waves. We will show below that this is in fact the case.

6.8.2. Forced, damped equatorial long wave theory

Let $\bar{\rho}$ and h be the average density and height of the frictional surface layer. Non-dimensionalize U and V with respect to $\bar{\rho}hc$, x and y with respect to $(c/\beta)^{1/2}(K_T/K)^{-1/4}$ and ϕ with respect to $\bar{\rho}hc^2(K_T/K)^{-1/2}$. Then the model equations (6.73) and (6.80) can be written in the non-dimensional form (Clarke 1994)

$$K'U' - y'V' = -\phi'_{x'} \tag{6.81}$$

$$K'V' + y'U' = -\phi'_y \tag{6.82}$$

$$K'\phi' + U'_{x'} + V'_y = -P' \tag{6.83}$$

where $'$ denotes a non-dimensional version of the original quantity,

$$K' = (\beta c)^{-1/2}(K_T)^{1/4}(K)^{3/4} \tag{6.84}$$

and

$$P' = aP(\bar{\rho}h)^{-1}(\beta c)^{-1/2}(K_T/K)^{-1/4}. \tag{6.85}$$

Note that for the realistic values $c = 68\,\mathrm{m\,s^{-1}}$, $\beta = 2.3 \times 10^{-11}\,\mathrm{m^{-1}\,s^{-1}}$, $K_T/K = 1/16$ and $K^{-1} = 2 \times 10^5\,\mathrm{s}$, the value of $K' \approx 0.06$.

For notational simplicity, drop the primes in (6.81)–(6.83). Then from (6.81) and (6.82),

$$(K^2 + y^2)U = -y\phi_y - K\phi_x. \tag{6.86}$$

Since $K \approx 0.06$ and the x scale of the precipitation forcing is much larger than the y scale, for most y

$$yU = -\phi_y. \tag{6.87}$$

In using (6.87) instead of (6.86) we clearly make an error over the small range $|y| \lesssim K$. Equation (6.87) has $y^2 U + y\phi_y = 0$ and (6.86) has

$$y^2 U + y\phi_y = -K\phi_x - K^2 U. \tag{6.88}$$

Since the RHS of (6.88) is tiny relative to typical values of $y^2 U$, the approximation $y^2 U + y\phi_y = 0$ (i.e., (6.87)) is reasonable.

Equations (6.87), (6.81) and (6.83) comprise a set of equations identical to the non-dimensional ocean set (5.1)–(5.3) except that the precipitation forcing P replaces the wind stress forcing (X, Y) and K replaces $\partial/\partial t$. By an exactly similar analysis to that of Section 5.2 one finds that the solutions for U, V and ϕ can be written in the form

$$U = \psi_0 q_0 + \sum_{m=1}^{\infty} q_m \left[\frac{\psi_{m+1}}{(m+1)^{1/2}} - \frac{\psi_{m-1}}{(m)^{1/2}} \right] \tag{6.89}$$

$$\phi = \psi_0 q_0 + \sum_{m=1}^{\infty} q_m \left[\frac{\psi_{m+1}}{(m+1)^{1/2}} + \frac{\psi_{m-1}}{(m)^{1/2}} \right] \tag{6.90}$$

$$V = \sum_{m=0}^{\infty} V_m \psi_m \tag{6.91}$$

where the $q_m(x)$ satisfy

$$K q_0 + q_{0x} = -\frac{1}{2}(P, \psi_0) \tag{6.92}$$

and

$$(2m+1) K q_m - q_{mx} = -\frac{1}{2}[(m+1)^{1/2} m(P, \psi_{m+1})$$
$$+ (m)^{1/2}(m+1)(P, \psi_{m-1})] \quad m = 1, 2, \ldots \tag{6.93}$$

and the V_m are given by

$$V_m = 2\sqrt{2} K q_m + \frac{(m+1)^{1/2}}{2^{1/2}}(P, \psi_{m+1}) + (m/2)^{1/2}(P, \psi_{m-1}) \quad m = 1, 2, \ldots \tag{6.94}$$

As anticipated, (6.89)–(6.94) are in the form of damped (K replaces $\partial/\partial t$) equatorial Kelvin and Rossby 'waves' forced by the precipitation. Putting $P = 0$ in (6.92) and (6.93) shows that the unforced equatorial

Kelvin wave decays exponentially eastward while the Rossby wave decays exponentially westward. The forcing term on the right-hand side of (6.93) can also be written as

$$\left(-\frac{1}{2}\right)(m)(m+1)\left[P, \frac{\psi_{m+1}}{\sqrt{m+1}} + \frac{\psi_{m-1}}{\sqrt{m}}\right],$$

i.e., the forcing for the mth damped Rossby wave uses the horizontal precipitation structure proportional to ϕ (see (6.90)). This makes sense physically because precipitation is proportional to vertical velocity and so is ϕ (see (6.78) and (6.79)). In the next section the forced wave physics will be used to explain why ENSO equatorial wind anomalies are westerly during warm ENSO episodes and easterly during cold ENSO episodes (Clarke 1994).

6.8.3. ENSO wind anomalies and forced wave physics

During warm ENSO episodes the western equatorial Pacific warm pool moves eastward and the deep atmospheric convective heating moves with it, causing anomalous deep convective heating in the equatorial region between approximately 150°E and 150°W (see Fig 2.2(c)). How should the atmosphere respond to this anomalous heating? First suppose that all the atmospheric response to this isolated heating is in the form of damped Rossby waves. East of the region of forcing no Rossby waves can exist because, if they did, they would grow instead of decay away from the forcing region. Therefore, east of the region of forcing $U = V = 0$. But in the region of heating, air must be rising out of the frictional boundary layer and consequently there must be horizontal convergence in the boundary layer. Part of this convergence is zonal. The only way to have zonal convergence and $U = 0$ at the eastern end of the forcing region is for $U > 0$, i.e., winds are westerly in regions of isolated heating if the response is totally due to Rossby waves.

The north–south scale of the OLR (and therefore precipitation) is about 8° of latitude (see Fig. 2.2(c)). The dimensional north–south scale of the equatorial Kelvin wave is $\sqrt{2c/\beta}\,(K_T/K)^{-1/4}$. For $c = 68\,\mathrm{m\,s^{-1}}$, $\beta = 2.3 \times 10^{-11}\,\mathrm{m^{-1}\,s^{-1}}$, $K_T/K = 1/16$, the Kelvin scale is $\approx 4900\,\mathrm{km}$. This is much larger than the north–south scale of the forcing and, because of the scale mismatch, the ENSO precipitation only weakly stimulates the equatorial Kelvin wave. Since the Rossby waves can exist at smaller north–south scales, the Rossby waves dominate the response to ENSO heating. Under these circumstances, by the earlier physical arguments, we expect the anomalous winds to be westerly in the heating region. When there is anomalous cooling, the air sinks, there is near-surface zonal divergence and, by the damped wave argument, surface equatorial flow is easterly. Thus the direction of ENSO wind anomalies, crucial to ENSO dynamics, can be explained by forced damped wave physics.

6.8.4. Caveat

The above model works well if the anomalous precipitation forcing (P' in (6.83)) is given everywhere. However, the model should not be used to deduce rainfall teleconnections because these depend on dynamics above the surface frictional boundary layer and the model is inadequate in the upper air. For example, during a warm ENSO episode, the central equatorial Pacific is anomalously wet, the air rises anomalously there and must descend somewhere. Observations show that the air descends preferentially in the western equatorial Pacific but the simple model fails because it has air descending gently everywhere outside the forcing region. However, the simple model *is* adequate if P' is *given* in the western equatorial Pacific—by the physics discussed earlier, descending air into the surface frictional layer on the scale of the anomalous precipitation drives surface easterlies (see, e.g., 2.9(c)).

In Chapter 5 we showed how the equatorial ocean responds to low-frequency wind forcing. In Chapter 6 we have considered the effect of SST on the atmospheric heating and the equatorial surface wind response. We have thus examined both pieces of the coupled ocean–atmosphere ENSO problem. In the next chapter we will consider some simple coupled ENSO models.

 REFERENCES

Clarke, A. J., 1994: Why are surface equatorial ENSO winds anomalously westerly under anomalous large-scale convection? *J. Clim.* **7**, 1623–1627.

Gadgil, S., P. V. Joseph, and N. V. Joshi, 1984: Ocean–atmosphere coupling over monsoon regions. *Nature* (London), **312**, 141–143.

Gill, A. E., 1980: Some simple solutions for heat-induced tropical circulation. *Q. J. R. Meteorol. Soc.*, **106**, 447–462.

Gill, A. E., and E. M. Rasmusson, 1983: The 1982–83 climate anomaly in the equatorial Pacific. *Nature*, **306**, 229–234.

Godfrey, J. S., M. Nunez, E. F. Bradley, P. A. Coppin, and E. J. Lindstrom, 1991: On the net surface heat flux into the Western Equatorial Pacific. *J. Geophys. Res.*, **96** (Suppl.), 3391–3400.

Graham, N. E., and T. P. Barnett, 1987: Sea surface temperature, surface wind divergence and convection over tropical oceans. *Science*, **238**, 657–659.

Horel, J. D., and J. M. Wallace, 1981: Planetary-scale atmospheric phenomena associated with the Southern Oscillation. *Mon. Wea. Rev.*, **109**, 813–829.

Krueger, A. F., and T. I. Gray, 1969: Long-term variations in equatorial circulation and rainfall. *Mon. Weather. Rev.*, **97**, 700–711.

Matsuno, T., 1966: Quasi-geostrophic motions in the equatorial areas. *J. Meteorol. Soc. Jpn., Ser. II*, **44**, 25–43.

Neelin, D., 1988: A simple model for surface stress and low level flow in the tropical troposphere driven by prescribed heating. *Q. J. R. Meteorol. Soc.*, **114**, 747–770.

Shu, L., and A. J. Clarke, 2002: Using an ocean model to examine ENSO dynamics. *J. Phys. Oceanogr.*, **32**, 903–923.

Sud, Y. C., G. K. Walker, and K.-M. Lau, 1999: Mechanisms regulating sea-surface temperatures and deep convection in the tropics. *Geophys. Res. Lett.* **26**, 1019–1022.

Waliser, D. E., N. E. Graham, and C. Gautier, 1993: Comparison of the highly reflective cloud and outgoing longwave radiation datasets for use in estimating tropical deep convection. *J. Clim.*, **6**, 331–353.

Wang, B., 1988: Dynamics of tropical low-frequency waves: An analysis of the moist Kelvin wave. *J. Atmos. Sci.*, **45**, 2051–2065.

Zhang, C., 1993: Large-scale variability of atmospheric deep convection in relation to sea surface temperature in the tropics. *J. Clim.*, **6**, 1898–1913.

CHAPTER SEVEN

ENSO COUPLED OCEAN–ATMOSPHERE MODELS

Contents

7.1. OVERVIEW

Over the last two decades an enormous number of coupled ocean–atmosphere ENSO models have been proposed to understand and predict ENSO. The models are of four main types. The simplest, usually involving one or two dependent variables, have been used to understand basic ENSO physics. Examples of such models will be given in Sections 7.2 and 7.3. More complicated are the 'intermediate coupled models' which usually involve a linear stratified ocean model (see (5.1)–(5.3)) coupled to a simple linear atmospheric model with governing equations in the form (6.81)–(6.83). The most complete dynamical ENSO model consists of an ocean general circulation model coupled to an atmospheric general circulation model. The advantage of the more complex models is that they usually make fewer approximations and assumptions; the disadvantage is that often they are so complex that their physics is not easily understood. The fourth main type of coupled model is the 'hybrid coupled model' in which a dynamical ocean model is coupled to a statistical atmospheric model.

Neelin et al. (1998), Stockdale et al. (1998) and Delecluse et al. (1998) have reviewed coupled ocean–atmosphere ENSO models. I will not attempt here to review all of the coupled ENSO models that have been proposed, but rather will focus on three dynamical ones. The first (Section 7.2), a simple warm pool displacement model of ENSO, is an example of the 'delayed oscillator' ENSO mechanism. In this mechanism some sort of coupled ocean–atmosphere instability grows until a delayed negative feedback stops and then reverses the instability. The reversal continues until a small perturbation of opposite sign arises and grows via the coupled ocean–atmosphere instability. This eventually is also reversed by the 'delayed negative feedback,' etc., so that an oscillation results. Delayed oscillator theory was first proposed by Schopf (1987) and was shown to be operating in more complicated coupled ocean–atmosphere models by Schopf and Suarez (1988), Suarez and Schopf (1988), Battisti (1988) and Battisti and Hirst (1989).

The second coupled ocean–atmosphere model to be discussed in detail (in Section 7.3) is the discharge–recharge oscillator of Jin (1997a,b). This model emphasizes the major role of the storage of equatorial heat and how that leads to a self-sustaining oscillation. Our treatment will differ from Jin's in that it emphasizes the key role played by the wind stress curl anomaly. Discharge–recharge oscillator physics is a basic component in some ENSO models (e.g., in the Cane and Zebiak 1985 model) and certain key relationships between SST, wind and upper ocean heat content that are predicted by the discharge–recharge oscillator have been observed (see Meinen and McPhaden, 2000 and Section 7.3).

The third coupled model to be discussed (in Section 7.4) is the intermediate coupled model of Cane and Zebiak (1985). This widely used and analyzed model was the first dynamical coupled model to predict El Niño. It has elements of both delayed oscillator and discharge–recharge physics.

In summary, this chapter focuses on examples of the two major ENSO oscillator paradigms (delayed oscillator (Section 7.2) and discharge–recharge oscillator (Section 7.3)) and the most widely used and studied coupled ENSO model. Initialization and the predictive capability of dynamical ENSO models will be discussed in Chapter 10.

7.2. Delayed Oscillator Theory

7.2.1. Introduction

Delayed oscillator theory often emphasizes SSTAs in the eastern equatorial Pacific, because in many models air–sea coupling is strongest there. But in fact observations suggest (see Chapter 2) that the main coupling occurs at the edge of the western equatorial Pacific warm pool. Model error occurs because, to a first approximation, the anomalous heating that typically takes place in the atmospheric component of the model depends primarily on the SSTAs which are biggest in the eastern equatorial Pacific. But the anomalous heating of the atmosphere depends not only on the SSTA but also on whether the total SST is high enough for deep atmospheric convection (see Sections 6.3–6.5); although SSTAs are largest in the eastern equatorial Pacific, the total SST is usually too low for deep atmospheric convection. Consequently, the most important SSTAs in ENSO dynamics are over the warm water in the western equatorial Pacific, particularly in the west-central equatorial Pacific at the edge of the western equatorial Pacific warm pool.

Figure 2.5 showed that the SOI and the zonal displacement of the eastern edge of the warm pool (28.5°C isotherm) are highly correlated. When the warm pool moves we expect the deep atmospheric convection and zonal winds it generates to move with it. Thus we expect the cloudiness at the edge of the warm pool (as measured by the OLR $240 \, W \, m^{-2}$ isoline) and the $-4 \, m \, s^{-1}$ eastward wind isotach to track closely with the 28.5°C isotherm. Figure 6.1 shows that these isolines do track each other quite closely. Due to the rain falling on it, the surface waters of the warm pool are fresher than waters to the east of it and consequently there is a sharp salinity front (Picaut et al. 1996). Observations (Picaut et al. 1996, 2001) show that this also closely tracks the other isolines in accordance with a zonal advection mechanism.

Because interannual displacement of the eastern edge of the western equatorial Pacific warm pool rather than eastern equatorial Pacific SSTA seems central to ENSO dynamics, rather than present a traditional version of the delayed oscillator model involving the eastern equatorial SSTA, I will focus on delayed oscillator theory for zonal equatorial displacements of the eastern edge of the western equatorial Pacific warm pool. Most elements of this model were originally described conceptually by Gill and Rasmusson (1983). Li (1996), Picaut et al. (1997) and Clarke et al. (2000) have all considered mathematical models of this type. We will follow the theory presented by Clarke et al. (2000).

7.2.2. Formulation of a warm pool displacement/delayed oscillator model of ENSO

We consider the zonal displacement of the eastern part of the equatorial western Pacific warm pool and base our model on the 28.5°C isotherm displacement index of Figs 2.5 and 6.1. Specifically, average the SST from 4°S to 4°N and let

$$x = x(t) \tag{7.1}$$

be the anomalous eastward displacement of the 28.5°C isotherm on the eastern side of the warm pool. Gill (1983) and more recently McPhaden and Picaut (1990) and Picaut and Delcroix (1995) suggest that the warm pool and 28.5°C isotherm displacement are mainly due to zonal advection by an anomalous zonal ocean current u'. Consistent with these results, Liu et al. (1994) found that in the central equatorial Pacific, anomalous net solar surface heat flux and anomalous evaporation are of minor importance to SST change. Recently Wang and McPhaden (2000) have shown that vertical advection and mixing also contribute to SST anomalies at the edge of the warm pool. However, to illustrate the warm pool displacement/delayed oscillator mechanism as simply as possible we assume that anomalous zonal advection is the only mechanism operating and write

$$\frac{dx}{dt} = u'(t). \tag{7.2}$$

The anomalous zonal ocean current u' consists of three parts which we denote u'_1, u'_2 and u'_3. When x is positive, the warm pool will be eastward of its normal position and anomalous deep atmospheric convection will generate westerly surface equatorial winds (see Section 6.8). These anomalous eastward winds generate an anomalous eastward local flow u'_1 and two delayed negative feedback flows u'_2 and u'_3. The local flow is essentially in phase with the anomalous wind stress and $x(t)$ and we write

$$u'_1(t) = ax(t). \tag{7.3}$$

If the anomalous zonal equatorial flow u' were entirely due to the local current u'_1, then the right-hand side of (7.2) would be given by (7.3) and the solution would be an exponentially growing solution. Physically, a small eastward displacement of the warm pool (small $x(t)$) will, as noted above, cause anomalous deep atmospheric convection, westerly wind anomalies and an eastward flow. This displaces the warm pool further to the east, generating increased deep atmospheric convection, a stronger westerly wind anomaly, an increased eastward ocean flow and further eastward warm pool displacement, etc.

The second current anomaly u_2' results from wave propagation. The anomalous westerly winds generate Rossby waves that propagate westward, reflect at the western ocean boundary and return as equatorial Kelvin waves (see Section 5.7). The Rossby waves have eastward currents and the reflected equatorial Kelvin waves have westward currents in order to satisfy the western equatorial boundary condition. When these equatorial Kelvin waves reach the eastern edge of the equatorial warm pool, the westward equatorial currents will try to return the 28.5°C isotherm to its original position, i.e., u_2' provides a negative feedback. Mathematically, since the currents are of opposite sign to the locally driven current u_1' we may write

$$u_2' = -bx(t - \Delta) \qquad (7.4)$$

where Δ is the time it takes for Rossby waves to propagate from the region of strong ocean–atmosphere coupling (near the date line) to the western boundary, reflect as an equatorial Kelvin wave and return to the region of strong ocean–atmosphere coupling. Both observations and theory (Li and Clarke 1994) suggest that it should take Rossby waves about 3 months to get to the western boundary and, theoretically, it should take the equatorial Kelvin waves 1 month to get back so $\Delta = 4$ months. But when the warm pool moves these travel times will change. Based on the result that the first vertical mode Kelvin wave (speed c) and the first vertical mode, first meridional mode Rossby wave (speed $c/3$) dominate the flow averaged over 4°S–4°N (Picaut and Delcroix 1995), we write

$$\Delta = 4 \text{ months} + \frac{3x(t - \Delta)}{c} + \frac{x(t)}{c}. \qquad (7.5)$$

Note that (7.5) defines a non-linear problem for Δ which is rapidly solved by iteration. At each time step t during the integration of (7.2) the initial guess for $\Delta(t)$ is Δ at the previous time step.

The third contribution to the current anomaly, u_3', is of similar form to the second. The second contribution involves generation of Rossby waves and reflection of these from the western Pacific Ocean boundary while the third contribution involves the generation of equatorial Kelvin waves and their reflection from the eastern Pacific boundary as Rossby waves. Both contributions are qualitatively the same, exerting a delayed negative feedback on the flow. Although solutions can be obtained when u_3' is included (Clarke et al. 2000), to keep the mathematics simple, we will ignore u'_3. Thus the governing equation (7.2), using (7.3) and (7.4), can be written as the non-linear equation

$$\frac{dx}{dt} = ax(t) - bx(t - \Delta). \tag{7.6}$$

7.2.3. Model solutions

We begin our analysis of (7.6) by considering the simple case when the delay Δ is constant. This equation has exactly the same form as the delayed oscillator model of Battisti and Hirst (1989) and, except for a cubic damping term, has the same form as the delayed oscillator model of Schopf (1987) and Suarez and Schopf (1988). These delayed oscillator models differ physically from (7.6) in that they are not based on warm pool displacement as the main component of ENSO physics.

Following Battisti and Hirst (1989), look for solutions of the $\Delta =$ constant linear equation (7.6) in the form

$$x(t) = A \exp(\sigma t) \cos \omega t \tag{7.7}$$

by substituting $A \exp(\sigma t + i\omega t)$ for $x(t)$ in (7.6). We obtain

$$\sigma + i\omega = a - b \exp[-(\sigma + i\omega)\Delta]. \tag{7.8}$$

Battisti and Hirst (1989) plotted and discussed solutions for σ and ω over a wide parameter range. Here our interest is in the low frequencies relevant to ENSO and we obtain analytical results (Clarke et al. 2000) for

$$\zeta = \omega\Delta \quad \text{less than or of the order of 1.} \tag{7.9}$$

To do this, multiply (7.8) by Δ and, after separating into real and imaginary parts, obtain

$$\xi = a\Delta - b\Delta \cos \zeta e^{-\xi} \tag{7.10}$$

$$\zeta = b\Delta \sin \zeta e^{-\xi} \tag{7.11}$$

where the two dimensionless unknowns are ζ (see (7.9)) and

$$\xi = \sigma\Delta. \tag{7.12}$$

From (7.11)

$$b\Delta e^{-\xi} = \frac{\zeta}{\sin \zeta} \tag{7.13}$$

and

$$\xi = \ln b\Delta + \ln \left(\frac{\sin \zeta}{\zeta} \right). \tag{7.14}$$

Substituting these results into (7.10) gives an equation for ζ:

$$\ln (b\Delta) + \ln \left(\frac{\sin \zeta}{\zeta} \right) = a\Delta - \frac{\zeta \cos \zeta}{\sin \zeta}. \tag{7.15}$$

For ζ less than or of the order of 1, with small error,

$$\ln \left(\frac{\sin \zeta}{\zeta} \right) = \ln \left(1 - \frac{\zeta^2}{6} \right) = -\frac{\zeta^2}{6} \tag{7.16}$$

and

$$\frac{\zeta \cos \zeta}{\sin \zeta} = 1 - \frac{\zeta^2}{3}. \tag{7.17}$$

Thus, from (7.15), (7.16) and (7.17) we obtain

$$\zeta \approx \sqrt{2} \left[\ln b\Delta - a\Delta + 1 \right]^{1/2} \tag{7.18}$$

and hence from (7.14), (7.16) and (7.18)

$$\xi = \frac{1}{3} (2 \ln b\Delta + a\Delta - 1). \tag{7.19}$$

It follows from (7.18) and $\zeta = \omega\Delta$ that the period T of the oscillation is

$$T = \sqrt{2}\pi \, \Delta / (\ln b\Delta - a\Delta + 1)^{1/2} \tag{7.20}$$

and from (7.19) and $\xi = \sigma\Delta$ that the e-folding growth rate time scale σ^{-1} is

$$\sigma^{-1} = 3\Delta / (2 \ln b\Delta + a\Delta - 1). \tag{7.21}$$

Numerical solutions of (7.6) show that the approximate analytical formulae, valid for small ζ^2, have small error even when $\zeta = 1$.

The formulae (7.20) and (7.21) indicate that the model oscillations can be of ENSO periodicity for reasonable parameter values. To show this, we must first estimate a and b. From (7.4) the equatorial Kelvin waves have currents

$$\sim bx \sim 0.2 \text{ m s}^{-1}. \tag{7.22}$$

The above estimate follows from the theory of Clarke (1991) which indicates that the sea level of the reflected equatorial Kelvin wave at the equator is comparable to the ENSO sea level signal seen on Australia's western coast. Since the latter amplitude is about 10 cm (see, e.g., Li and Clarke 2004), the reflected equatorial Kelvin wave sea level amplitude at the equator is also about 10 cm. This corresponds to a zonal surface current, averaged between 4°S and 4°N, of about 20 cm s^{-1} (see 7.22). Since typical amplitudes for x are $\sim$2000 km, for $\Delta = 4$ months

$$b\Delta \sim 1. \tag{7.23}$$

But the Rossby wave mass flux onto the western boundary, generated remotely through the ENSO air–sea interaction, is comparable to the reflected equatorial Kelvin wave mass flux (Clarke 1991; du Penhoat and Cane 1991). Therefore we have

$$ax \sim bx, \tag{7.24}$$

and hence

$$b\Delta \sim a\Delta \sim 1. \tag{7.25}$$

For such a parameter range, (7.20) and (7.21) show that growing, neutral or decaying long-period solutions are possible.

Notice that the period T of the coupled oscillation (see (7.20)) depends both on the wave transit time Δ and the time scales a^{-1} and b^{-1} associated with the anomalous advecting flow. Physically, if the negative feedback is sufficiently delayed, there is time for the instability to grow before being damped out by the negative feedback. There is also time, once the damping has overpowered the instability, for the negative feedback to change the sign of $x(t)$, e.g., from positive to negative. Mathematically, in (7.6) when $x(t) = 0$ and $x(t - \Delta) > 0$, the negative feedback $-bx(t - \Delta) < 0$ so that $dx/dt < 0$ and a small negative $x(t)$ subsequently results. This small negative perturbation can then grow and be restrained by delayed negative feedback so that eventually a small positive (eastward) perturbation of the warm pool results. In this way a sequence of warm and cold ENSO episodes occurs. The oscillation periodicity, which is about 2–7 years, is much longer than the 4-month wave transit delay time; although the delay time is crucial to the existence of the period of the oscillation, the oscillation period also depends critically on the relative importance of the growth rate of the instability and the negative feedback.

7.2.4. The influence of other negative feedbacks

So far we have only considered one negative feedback but there are others. One is associated with the easterly (westerly) wind anomalies which form in the far western equatorial Pacific during warm (cold) ENSO episodes, particularly during December–February (see Fig. 2.9 and Weisberg and Wang 1997). These far western equatorial Pacific wind anomalies generate equatorial Kelvin waves with zonal flow opposite to the flow of the growing instability near the date line. Since the equatorial Kelvin waves only take about 1 month to propagate from the western equatorial Pacific to the date line, this negative feedback is essentially in phase with $-x(t)$. Consequently, its main effect will be to reduce the parameter a in the growth term $ax(t)$. This effect is likely to be small or negligible, since calculations by Shu and Clarke (2002) suggest that this wind feedback is ineffective.

Another negative feedback is due to the surface heat flux. Specifically, increased eastward warm pool displacement implies increased deep atmospheric convection, more clouds, more shade, less heating by the sun's incoming short wave radiation and hence cooling, i.e., the original SST anomaly tends to be damped out (McPhaden and Hayes 1991; Ramanathan and Collins 1991; Waliser et al. 1994). Similar negative feedback arguments apply for a negative SST anomaly. The effect of anomalous cloud cover is approximately proportional to $-x(t)$ and so will reduce the parameter a in the growth term $ax(t)$.

In terms of our simple model, overall the negative feedbacks result in oscillations which do not grow or decay. For this $\sigma = 0$ case, by (7.19), our simple model has

$$1 - a\Delta = 2 \ln b\Delta \qquad (7.26)$$

and hence, by (7.20), the period T of the oscillation is

$$T = \sqrt{2}\pi\Delta/(3 \ln b\Delta)^{1/2}. \qquad (7.27)$$

For $\Delta = 4$ months,

$$T = 10.3/(\ln b\Delta)^{1/2} \text{ months.} \qquad (7.28)$$

Therefore, for interannual periodicity and zero growth rate, $\ln b\Delta$ must be a small positive number, i.e., $b\Delta$ must be restricted to a number slightly greater than 1. It follows from (7.21) that $a\Delta$ must be nearly 1 for neutrally stable oscillations. This makes sense physically. If $a\Delta$ were much greater than 1, $x(t)$ in (7.6) would grow too much before the negative feedback could control it and produce a stable oscillation; if $a\Delta$ were small $x(t)$ would take much longer than Δ to grow or decay so Δ would be negligible in (7.6) and decaying $(b > a)$ or growing $(a > b)$ solutions would result.

One weakness in the theory is that for realistic interannual periodicity, parameter values must occur only over a narrow range. For example, if we want periods between say 2 and 7 years, then from (7.28)

$$1.015 \leq b\Delta \leq 1.202 \tag{7.29a}$$

and by (7.26)

$$0.6320 \leq a\Delta \leq 0.9702. \tag{7.29b}$$

7.2.5. Non-constant Δ

So far the delay Δ has been taken to be constant rather than the more accurate time-varying x-dependent value in (7.5). Calculations (see Clarke et al. 2000) show that variable Δ does not affect the solution of (7.7) very much. We might expect a bigger effect when, as in more realistic ENSO time series, the amplitude of $x(t)$ changes considerably from one cycle to the next.

One way ENSO irregularity can arise is via random wind-forced currents affecting the eastward displacement of the warm pool. For simplicity, we introduce such irregularity into our model by adding a white noise current $\varepsilon(t)$ to the right-hand side of (7.6) so that it becomes

$$\frac{dx}{dt} = ax(t) - bx(t - \Delta) + \varepsilon(t). \tag{7.30}$$

When $a = b = 0$, the solution of (7.25) is 'red noise,' i.e., a solution with increasing amplitude at lower frequencies. When a and b are non–zero and Δ is constant, the solution $x(t)$ has variable amplitude but the period T is nearly constant and corresponds to (7.20). More realistically, when Δ depends on x, the simple model $x(t)$ is irregular in both amplitude and T but with dominant interannual periodicity.

7.3. Discharge–Recharge Oscillator Theory

7.3.1. Introduction

Over the last several years the two most popular paradigms for the basic ENSO mechanism have been the delayed oscillator mechanism, a version of which has just been given in Section 7.2, and the discharge–recharge oscillator mechanism presented by Jin (1997a,b). Here we will present a version of the discharge–recharge oscillator following Clarke et al. (2007). This model is similar to Jin's in that it emphasizes the major role of equatorial heat storage but differs from Jin's in that it emphasizes the wind stress curl anomalies and the central rather than eastern equatorial Pacific SST anomalies.

Our model focuses on the 5°S–5°N equatorial strip in the Pacific. The goal is to illustrate simply the main physical elements in the ocean–atmosphere coupling and how oscillations can occur. Necessarily, therefore, several approximations will be made and our model will consist of only two central equatorial variables: T_{cen}, the SST anomaly averaged over the region 5°S–5°N, 170°W–150°W, and D_{cen}, the 20°C isotherm depth anomaly averaged over the same region. The coupled model will consist of two parts, one in which the atmosphere drives ocean (Section 7.3.2) and the other in which ocean dynamics and thermodynamics change the SST which then drives the atmosphere (Section 7.3.3).

7.3.2. The atmosphere drives the ocean

One component of the model is built directly on the theoretical and observational results of Section 5.6. Equation (5.61) tells us that the anomalous negative curl forcing is proportional to $(\partial/\partial t)(\int_S D \, dS)$. As noted in Chapter 5, consistent with a rapid response of the atmosphere to SST, the negative wind stress curl forcing is proportional to NINO3.4. Since NINO3.4 and T_{cen} are highly correlated ($r = 0.96$) we may write

$$\frac{\partial}{\partial t}\left(\int_S D \, dS\right) \approx (\text{negative constant})\, T_{cen}. \qquad (7.31)$$

To relate $\int_S D \, dS$ to D_{cen} recall that in Chapter 5 we showed that the interannual thermocline variability is dominated by two empirical orthogonal function (EOF) modes; the first mode (Fig. 5.3a) changes sign at 160°W whereas the second mode (Fig. 5.5a) is of one sign across the equatorial Pacific. Therefore D_{cen}, the average of D from 170°W to 150°W, has no contribution from the first EOF mode and essentially all its contributions are from the second EOF mode. But we also found in Chapter 5 that $\int_S D \, dS$ is proportional to the second EOF so D_{cen} and $\int_S D \, dS$ are proportional too. Hence we may substitute D_{cen} for $\int_S D \, dS$ in (7.31) to get

$$\frac{\partial D_{cen}}{\partial t} = -\mu T_{cen} \tag{7.32}$$

for some positive constant μ. Correlation of interannual $\partial D_{cen}/\partial t$ with interannual T_{cen} supports the approximate relationship (7.32) since the correlation coefficient is quite negative ($r = -0.76$). The regression coefficient $\mu = 17.2 \, \mathrm{m} \, (\mathrm{year})^{-1} \, (°\mathrm{C})^{-1}$.

7.3.3. Ocean dynamics and thermodynamics drive T_{cen} and the atmosphere

Equation (7.32) describes how T_{cen} drives interannual fluctuations in D_{cen}. But it is also true that D_{cen} can drive interannual variations in T_{cen}. Consider, for example, an idealized case when the ocean mixed layer has a constant depth and water is stratified beneath the mixed layer. When the thermocline depth anomaly $D_{cen} = 0$, normal wind-generated turbulence at the base of the mixed layer results in cooler water parcels moving into the mixed layer and warmer parcels leaving it, i.e., there is a heat flux out of the mixed layer but no net exchange of mass. This mean heat flux is balanced by mean heat fluxes into the mixed layer from the surface and/or horizontal advection so that the anomalous mixed layer temperature, which is the same as the anomalous SST T_{cen}, is zero. Now suppose that $D_{cen} < 0$, i.e., the 20°C isotherm is displaced anomalously upward. Since the mixed layer depth is constant, conservation of mass requires that in the non-mixing region above the 20°C isotherm, water must have diverged horizontally. There is thus cooler water at the base of the mixed layer and so, even though the wind stress may be unaltered, wind-generated turbulence at the base of the mixed layer causes an anomalous heat flux down through the bottom of the mixed layer. The anomalous heat flux will cause $\partial T_{cen}/\partial t$ to be negative. The more negative the anomalous 20°C isotherm displacement, the bigger the negative heat flux anomaly and the more negative $\partial T_{cen}/\partial t$. Similar arguments suggest that if $D_{cen} > 0$, then $\partial T_{cen}/\partial t > 0$. Based on these arguments, we expect

$$\frac{\partial T_{cen}}{\partial t} = \nu D_{cen} \tag{7.33}$$

for some positive constant ν. Observed interannual $\partial T_{\text{cen}}/\partial t$ and D_{cen} support the relationship (7.33) since their time series are quite well correlated ($r = 0.68$). The regression coefficient ν is $0.099°\text{C} (\text{year})^{-1} (\text{m})^{-1}$.

The relationship (7.33) is also consistent (Jin and An 1999) with advection of the mean SST $\overline{T}$ by the anomalous flow

$$\frac{\partial T_{\text{cen}}}{\partial t} = -u'\frac{\partial \overline{T}}{\partial x} \tag{7.34}$$

provided that u' is proportional to D_{cen}. Jin and An (1999) argue for this proportionality because u' is related to the meridional gradient of the thermocline via geostrophic balance. Observations by Wang and McPhaden (2000) suggest that both zonal advection and entrainment through the base of the mixed layer contribute to SST anomalies in the central equatorial Pacific.

7.3.4. Mathematical solution and coupled physics

Differentiation of (7.34) with respect to time and substituting for D_{cen} using (7.33) gives

$$\frac{\partial^2}{\partial t^2}(T_{\text{cen}}) + \omega^2 T_{\text{cen}} = 0 \tag{7.35}$$

with

$$\omega = \sqrt{\mu\nu}. \tag{7.36}$$

For the parameter values $\mu = 17.2\,\text{m} (\text{year})^{-1} (°\text{C})^{-1}$ and $\nu = 0.099°\text{C} (\text{year})^{-1} (\text{m})^{-1}$, determined earlier by regression (see (7.32) and (7.33))

$$\omega = 2\pi/4.8\,\text{years}, \tag{7.37}$$

a reasonable interannual frequency.

The physics of the oscillation is summarized in Fig. 7.1. We begin our discussion of the oscillation at the height of an El Niño (first panel of the figure). Then T_{cen} is maximally positive, there is anomalous equatorial deep atmospheric convection and the wind anomalies are maximally eastward. The wind anomalies cause a twofold ocean response (see Sections 5.4–5.6). One part of the response consists of a tilt of the thermocline in phase with the westerly winds as the wind stress forcing is in quasi-steady balance with the zonal pressure gradient. While this tilt affects the thermocline depth in the eastern and western equatorial Pacific, the displacement in

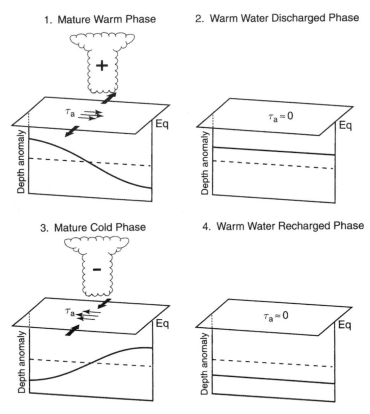

Fig. 7.1 An idealized schematic of the discharge–recharge oscillator. All quantities shown are anomalies. The dashed and heavy solid lines denote the zero thermocline depth anomaly and the actual thermocline depth anomaly, respectively. The anomalous heating and cooling of the atmosphere occurs in the west-central equatorial Pacific through anomalous deep atmospheric convection denoted by the cloud symbols, the plus and minus notations denoting positive and negative anomalies. The atmospheric convection anomalies are driven by west-central equatorial Pacific SST anomalies of the same sign (not shown). The thin arrows at the surface refer to the zonal equatorial wind anomalies with stress τ_a. These anomalies, driven by the deep anomalous atmospheric convection, decrease meridionally with latitude and the resultant wind stress curl anomalies drive meridional ocean transport (solid thick arrows). The text discusses the oscillation in the panel sequence 1–4. (Based on Clarke et al. 2007).

the central equatorial Pacific is nearly zero. The second part of the response is due to the anomalous wind stress curl which causes a meridional poleward transport of warm water (mathematically, when T_{cen} is a maximum the rate of loss of equatorial warm water is greatest in (7.32)). One quarter of a period later (second panel of the figure), the warm water has been discharged from the equator and the thermocline is anomalously shallow ($D_{cen} < 0$). A raised thermocline implies colder water nearer the surface and an anomalous negative heat flux through the base of the mixed layer so that T_{cen} decreases (see (7.33)). It also results in a westward equatorial current anomaly which advects cold water from the eastern equatorial Pacific, again resulting in T_{cen} decreasing (see (7.34)). Eventually, one quarter of a period later (third panel of the figure) T_{cen} has reached its negative extremum and the model exhibits La Niña conditions with less than usual deep atmospheric convection, an anomalous thermocline tilt and equatorial heat content anomaly zero. Through their curl, the easterly equatorial wind anomalies cause a transport of water onto the equator so that a quarter period later (see panel 4 of the figure) the warm water volume on the equator is at its maximum ($D_{cen} > 0$). This then results in a positive heat flux anomaly at the base of the mixed layer and eastward current anomalies, causing $\partial T_{cen}/\partial t$ to be greater than zero (see (7.33)) and El Niño conditions to return a quarter of a period later.

The above idealized model errs in several respects. For example, observations (Meinen and McPhaden 2000) suggest that the same amplitude positive and negative warm water volume anomalies produce different amplitude SST anomalies whereas the idealized model makes no such distinction in the equivalent variables D_{cen} and T_{cen}. Kessler (2002) also found that the recharge–discharge oscillator only operates for three quarters of the cycle; there is a break in the cycle from recharge to El Niño. In addition, the idealized oscillator is not phase-locked to the seasonal cycle (see Chapter 8). Nevertheless, the idealized model does illustrate the idea that wind stress curl anomalies play a key role in ENSO dynamics and shows how they do this.

7.4. INTERMEDIATE COUPLED MODELS

The first coupled models to use a linear ocean model like that in Chapter 5 were the models of McCreary (1983) and McCreary and Anderson (1984). In these models the model atmosphere had a fixed horizontal structure and was coupled to the ocean through the ocean thermocline depth in the far eastern equatorial Pacific. Shortly after this, Cane and Zebiak (1985) (see also Zebiak and Cane (1987)) constructed an intermediate coupled model which not only has been widely used to understand El Niño dynamics, but also was the first dynamical model to predict El Niño.

The atmospheric component of the Cane–Zebiak model is based on Gill (1980) and is similar to the steady-state linear atmospheric model in Chapter 6 (see (6.81)–(6.83)). The model atmosphere is driven by anomalous heating Q' of the form

$$Q' = Q'_S + Q'_C \tag{7.38a}$$

where

$$Q'_S = \alpha_* T' \exp\left\{\overline{T} - 30°C/16.7°C\right\} \tag{7.38b}$$

and

$$Q'_C = \beta_* \exp\left\{M\left(\overline{C} + C'\right) - M(\overline{C})\right\}. \tag{7.38c}$$

In (7.38) $\overline{T}$ is the mean SST, T' the anomalous SST, α_* and β_* are constants, $\overline{C}$ is the mean surface wind horizontal convergence and the function

$$\begin{aligned} M(\xi) &= \xi \quad \text{for } \xi > 0 \\ &= 0 \quad \text{otherwise.} \end{aligned} \tag{7.39}$$

The anomalous heating thus depends on the SST anomaly T' and non-linearly on the mean and total atmospheric convergence. The model is an iterative one; the SST anomaly gives rise to an nth step total anomalous heating Q'_n which generates anomalous surface winds, a new surface convergence anomaly C'_n and an $(n+1)$th step total anomalous heating Q'_{n+1}. Zebiak (1986) has discussed the convergence of this model.

The ocean component of the coupled model is a linear stratified wind-forced model like that of Chapter 5 (see (5.1)–(5.3)) with the stratification approximated by two layers of water of constant density and the dynamics governed by a single baroclinic mode. Wind anomalies from the atmospheric model drive anomalous horizontal ocean currents and upwelling which then change the SST through horizontal advection, vertical entrainment and anomalous heat fluxes in parameterized mixed layer physics. The SST anomaly produced then drives the atmospheric model which then produces wind anomalies to force the ocean and so on.

While the Cane–Zebiak model is an admirable early coupled ENSO model, it does have some deficiencies. Probably the biggest of these is its anomalous heating parameterization, which depends strongly on the sea surface temperature anomaly rather than on the SST anomaly over waters warmer than about 28°C where deep atmospheric convection occurs (see Chapter 6). Consequently, instead of the anomalous equatorial heating occurring near the edge of the warm pool (see Figs 2.2c, 2.4c and 2.9c), it occurs much further to the east (Dewitte and Perigaud 1996) where the SST anomalies are larger (but ride upon cold eastern equatorial Pacific waters).

Battisti (1988) constructed a near copy of the Cane and Zebiak model and examined its dynamics in detail. A key component of the mechanism is a delayed negative feedback via reflection of equatorial Rossby waves from the western equatorial Pacific boundary. This delayed negative feedback idea, independently found by Battisti (1988) but first described by Schopf (1987), has been discussed in Section 7.2. The delayed oscillator model of Battisti and Hirst (1989), a simplification of the Cane and Zebiak model, is governed by an equation identical in form to (7.6) except that the equatorial warm pool displacement $x(t)$ is replaced by the eastern equatorial Pacific SST anomaly. This is in keeping with the emphasis on eastern equatorial Pacific heating in the Cane and Zebiak model.

An intermediate coupled model which better takes into account the dependence of anomalous deep convective heating on total SST is that due to Kleeman (1993). Like the Cane and Zebiak model, oceanic wave reflection at the western Pacific boundary is crucial to the model.

While the delayed oscillator mechanism has been associated with the Cane and Zebiak model, Cane and Zebiak (1985), Zebiak and Cane (1987) and Zebiak (1989) have emphasized the importance of the model equatorial heat content. Specifically, Zebiak and Cane (1987) report that in model experiments 'where the effects of the heat content anomalies were artificially suppressed, the ENSO cycle was eliminated.' In other words, elements of both the delayed oscillator (Section 7.2) and the discharge–recharge oscillator (Section 7.3) seem to play a role in the Cane and Zebiak ENSO model.

7.5. CONCLUSION

In this chapter we have analyzed some ways in which the equatorial Pacific Ocean and atmosphere may couple together to produce an inter-annual oscillation. The delayed oscillator mechanism of Section 7.2 requires a delicate balance between the growth of a coupled ocean–atmosphere instability and a delayed negative feedback via equatorial waves. Observational evidence for and against this mechanism has been given (see, e.g., Delcroix et al. 2000). There is also observational support for the discharge–recharge oscillator theory of Section 7.3 (see, e.g., Meinen and McPhaden 2000; Clarke et al. 2007).

A mysterious ENSO characteristic was not addressed in this chapter. Specifically, why do ENSO indices, which have the seasonal cycle removed from their time series, nevertheless have properties strongly phase-locked to the seasonal cycle? For example, why should the July time series of the anomaly index NINO3.4 be correlated with the following January time series at 0.85 but across the spring the January time series is correlated with the following July at only 0.03? This striking seasonally phase-locked persistence, fundamental to ENSO dynamics and prediction and known since the work of Walker (1924) in colonial India, will be discussed in the next chapter.

> ## REFERENCES

Battisti, D. S., 1988: The dynamics and thermodynamics of a warming event in a coupled tropical atmosphere–ocean model. *J. Atmos. Sci.*, **45**, 2889–2919.

Battisti, D. S., and A. C. Hirst, 1989: Interannual variability in a tropical atmosphere–ocean model: Influence of the basic state, ocean geometry and nonlinearity. *J. Atmos. Sci.*, **46**, 1687–1712.

Cane, M. A., and S. E. Zebiak, 1985: A theory for El Niño and Southern Oscillation. *Science*, **228**, 1084–1087.

Clarke, A. J., 1991: On the reflection and transmission of low-frequency energy at the irregular western Pacific Ocean boundary. *J. Geophys. Res.*, **96**, 3289–3305.

Clarke, A. J., S. Van Gorder, and G. Colantuono, 2007: Wind stress curl and ENSO discharge/recharge in the equatorial Pacific. *J. Phys. Oceanogr.*, **37**(4), 1077–1091.

Clarke, A. J., J. Wang, and S. Van Gorder, 2000: A simple warm-pool displacement ENSO model. *J. Phys. Oceanogr.*, **30**(7), 1679–1691; Corrigendum *JPO*, **30**(12), 3271.

Delcroix, T., B. Dewitte, Y. duPenhoat, F. Masia, and J. Picaut, 2000: Equatorial waves and warm pool displacements during the 1992–1998 El Niño Southern Oscillation events: Observation and modeling. *J. Geophys. Res.*, **105**, 26,045–26,062.

Delecluse, P., M. K. Davey, Y. Kitamura, S. G. H. Philander, M. Suarez, and L. Bengtsson, 1998: Coupled general circulation modeling of the tropical Pacific. *J. Geophys. Res.*, **103**, 14357–14373.

Dewitte, B., and C. Perigaud, 1996: El Niño–La Niña events simulated with Cane and Zebiak's model and observed with satellite or in situ data. Part II: Model forced with observations. *J. Clim.*, **9**, 1188–1207.

du Penhoat, Y., and M. A. Cane, 1991: Effect of low-latitude western boundary gaps on the reflection of equatorial motions. *J. Geophys. Res.*, **96** (Suppl.), 3307–3322.

Gill, A. E., 1980: Some simple solutions for heat-induced tropical circulation. *Q. J. R. Meteorol. Soc.*, **106**, 447–462.

Gill, A. E., 1983: An estimation of sea level and surface-current anomalies during the 1972 El Niño and consequent thermal effects. *J. Phys. Oceanogr.*, **13**, 586–606.

Gill, A. E., and E. M. Rasmusson, 1983: The 1982–83 climate anomaly in the equatorial Pacific. *Nature*, **306**, 229–234.

Jin, F.-F., 1997a: An equatorial ocean recharge paradigm for ENSO. Part I: Conceptual model. *J. Atmos. Sci.*, **54**, 811–829.

Jin, F.-F., 1997b: An equatorial ocean recharge paradigm for ENSO. Part II: A stripped-down coupled model. *J. Atmos. Sci.*, **54**, 830–847.

Jin, F.-F., and S-I. An, 1999: Thermocline and zonal advective feedbacks within the Equatorial Ocean Recharge Oscillator Model for ENSO. *Geophys. Res. Lett.*, **26**, 2989–2992.

Kessler, W. S., 2002: Is ENSO a cycle or a series of events? *Geophys. Res. Lett.*, **29**(23), 2125, doi:10.1029/2002GL015924.

Kleeman, R., 1993: On the dependence of hindcast skill on ocean thermodynamics in a coupled ocean–atmosphere model. *J. Clim.*, **6**, 2012–2033.

Li, B., 1996: On the timing of warm and cold El Niño/Southern Oscillation events. PhD dissertation, Florida State University, 64 pp.

Li, B., and A. J. Clarke, 1994: An examination of some ENSO mechanisms using interannual sea level at the eastern and western equatorial boundaries and the zonally averaged equatorial wind. *J. Phys. Oceanogr.*, **24**, 681–690.

Li, J., and A. J. Clarke, 2004: Coastline direction, interannual flow and the strong El Niño currents along Australia's nearly zonal southern coast. *J. Phys. Oceanogr.*, **34**(11), 2373–2381.

Liu, W. T., A. Zhang, and J. K. B. Bishop, 1994: Evaporation and solar irradiance as regulators of sea surface temperature in annual and interannual changes. *J. Geophys. Res.*, **99**, 12,623–12,637.

McCreary, J. P., 1983: A model of tropical ocean-atmospheric interaction. *Mon. Weather Rev.*, **111**, 370–387.

McCreary, J. P., and D. L. T. Anderson, 1984: A simple model of El Niño and the Southern Oscillation. *Mon. Weather Rev.*, **112**, 934–946.

McPhaden, M. J., and S. P. Hayes, 1991: On the variability of winds, sea surface temperature, and surface layer heat content in the western equatorial Pacific. *J. Geophys. Res.*, **96**(Suppl.), 3331–3342.

McPhaden, M. J., and J. Picaut, 1990: El Niño–Southern Oscillation displacement of the western equatorial Pacific warm pool. *Science*, **250**, 1385–1388.

Meinen, C. S., and M. J. McPhaden, 2000: Observations of warm water volume changes in the equatorial Pacific and their relationship to El Niño and La Niña. *J. Clim.*, **13**, 3551–3559.

Neelin, J. D., D. S. Battisti, A. C. Hirst, F.-F. Jin, Y. Wakata, T. Yamagata, and S. E. Zebiak, 1998: ENSO theory. *J. Geophys. Res.*, **103**, 14261–14290.

Picaut, J., and T. Delcroix, 1995: Equatorial wave sequence associated with warm pool displacements during the 1986–1989 El Niño–La Niña. *J. Geophys. Res.*, **100**, 18393–18408.

Picaut, J., M. Ioualalen, T. Delcroix, F. Masia, R. Murtugudde, and J. Vialard, 2001: The oceanic zone of convergence on the eastern edge of the Pacific warm pool: A synthesis of results and implications for El Niño–Southern Oscillation and biogeochemical phenomena. *J. Geophys. Res.*, **106**, 2363–2386.

Picaut, J., M. Ioualalen, C. Menkes, T. Delcroix, and M. J. McPhaden, 1996: Mechanism of the zonal displacements of the Pacific warm pool: Implications for ENSO. *Science*, **274**, 1486–1489.

Picaut, J., F. Masia, and Y. du Penhoat, 1997: An advective–reflective conceptual model for the oscillatory nature of the ENSO. *Science*, **277**, 663–666.

Ramanathan, V., and W. Collins, 1991: Thermodynamic regulation of ocean warming by cirrus clouds deduced from observations of the 1987 El Niño. *Nature*, **351**, 27–32.

Schopf, P. S., 1987: Coupled dynamics of the tropical ocean–atmosphere system. U.S. TOGA Workshop on the Dynamics of the Equatorial Oceans, Honolulu, Hawaii, TOGA, 279–286.

Schopf, P. S., and M. J. Suarez, 1988: Vacillations in a coupled ocean–atmosphere model. *J. Atmos. Sci.*, **45**, 549–567.

Shu, L., and A. J. Clarke, 2002: Using an ocean model to examine ENSO dynamics. *J. Phys. Oceanogr.*, **32**(3), 903–923.

Stockdale, T. N., A. J. Busalacchi, D. E. Harrison, and R. Seager, 1998: Ocean modeling for ENSO. *J. Geophys. Res.*, **103**, 14325–14355.

Suarez, M. J., and P. S. Schopf, 1988: A delayed action oscillator for ENSO. *J. Atmos. Sci.*, **45**, 3283–3287.

Waliser, D. E., B. Blanke, J. D. Neelin, and C. Gautier, 1994: Shortwave feedbacks and El Niño–Southern Oscillation: Forced ocean and coupled ocean-atmosphere experiments. *J. Geophys. Res.*, **99**, 25,109–25,125.

Walker, G. T., 1924: Correlation in seasonal variations in weather IX: A further study of world–weather. *Mem. Indian Meteorol. Dept*, **24**, 275–332.

Wang, W., and M. J. McPhaden, 2000: The surface-layer heat balance in the equatorial Pacific Ocean. Part II. Interannual variability. *J. Phys. Oceanogr.*, **30**, 2989–3008.

Weisberg, R. H., and C. Wang, 1997: A western Pacific oscillator paradigm for the El Niño–Southern Oscillation. *Geophys. Res. Lett.*, **24**, 779–782.

Zebiak, S. E., 1986: Atmospheric convergence feedback in a simple model for El Niño. *Mon. Weather Rev.*, **114**, 1263–1271.

Zebiak, S. E., 1989: Oceanic heat content variability and El Niño cycles. *J. Phys. Oceanogr.*, **19**, 475–486.

Zebiak, S. E., and M. A. Cane, 1987: A model El Niño–Southern Oscillation. *Mon. Weather Rev.*, **115**, 2262–2278.

CHAPTER EIGHT

PHASE-LOCKING OF ENSO TO THE CALENDAR YEAR

Contents

8.1. OVERVIEW

One of the mysterious and fundamental properties of ENSO is that, even after the annual cycle has been removed from the ENSO time series, the anomalous time series still exhibits properties phase-locked to the calendar year. For example, the Southern Oscillation and El Niño are persistent during the last half of the calendar year and early part of the next but not across the boreal spring. Figures 2.13 and 2.14 show this spring persistence barrier for the SOI and the El Niño index NINO3.4. The signal to noise ratio in NINO3.4 is high and the drop in persistence across April/May is sudden and large. Although the phase-locking of ENSO anomalies to the seasonal cycle is fundamental to ENSO dynamics, it is only just beginning to be understood.

The next section discusses observations of the ENSO calendar year phase-locking of SST and wind anomalies during warm ENSO events. Following this, in Section 8.3, we will examine the phase-locked structure in the NINO3.4 and SOI time series. The April persistence barrier in these indices suggests that random variability dominates at this time of the year and that therefore it may be impossible to predict past April. But examination of the zonal equatorial wind anomalies (Section 8.4) shows that the future ENSO signal, phase-locked to the calendar year, can be seen developing in the months before April. The dynamical link between the phase-locked zonal wind anomalies and the phase-locked structure in NINO3.4 and the SOI is discussed in Section 8.5. Many researchers have examined quasi-biennial variability in ENSO – the tendency of one phase-locked year to be followed by another of opposite sign. The biennial tendency is weak, however, and no widely accepted theory is available so this topic is treated only briefly in Section 8.6. A phase-locking mechanism for the termination of ENSO events is discussed in Section 8.7 and elements of a seasonally phase-locked ENSO mechanism in Section 8.8. Concluding remarks follow in Section 8.9.

8.2. The Composite Warm ENSO Event

Rasmusson and Carpenter (1982) averaged tropical Pacific data for the six El Niño episodes in 1951–1972 to produce a 'composite' view of ENSO. The data suggested that El Niño episodes are often similarly phase-locked to the calendar year. Figure 8.1 shows the degree to which El Niño SST anomalies in the eastern equatorial Pacific are phase-locked to the calendar year since 1950. Year (0), the El Niño year, is the year in which the El Niño develops. Based on the definition used by Rasmusson and Carpenter (1982) and Harrison and Larkin (1998), there have been 11 El Niño years (1951, 1953, 1957, 1965, 1969, 1972, 1976, 1982, 1987, 1991 and 1997) between 1950 and 2001. Figure 8.1 shows that the average anomaly (composite) is larger for the five El Niños post 1975 because of the huge 1982 and 1997 El Niños, but the calendar phase-locking is approximately the same for the El Niños pre and post 1975, with a maximum anomaly occurring near the end of the year. Similar phase-locking is evident for La Niña (see Torrence and Webster 1998 and Fig. 8.2).

Just as in Fig. 8.1 where we composite the El Niño index NINO3.4 into year (−1), year (0) and year (+1) for an El Niño episode, so gridded

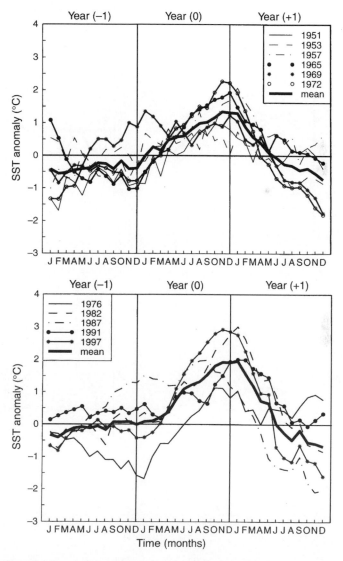

Fig. 8.1 The El Niño index NINO3.4 plotted for the El Niño year (year (0)), the year before the El Niño year (year (−1)) and the year after the El Niño year (year (+1)). The top panel shows the El Niño events and their mean NINO3.4 for the period 1950–1975 and the bottom panel the El Niño events and their mean NINO3.4 for the period 1976–1998.

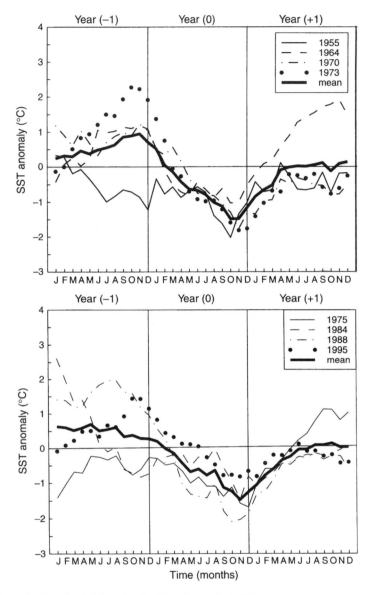

Fig. 8.2 As for Fig. 8.1 but for La Niña instead of El Niño.

SST, zonal and meridional wind anomalies can also be similarly composited. Harrison and Larkin (1998) have done this for the 10 El Niños during 1950–1994. Figures 8.3–8.5 show the composite results for the 18 months from January of year (0) to June of year (+1). This time interval corresponds

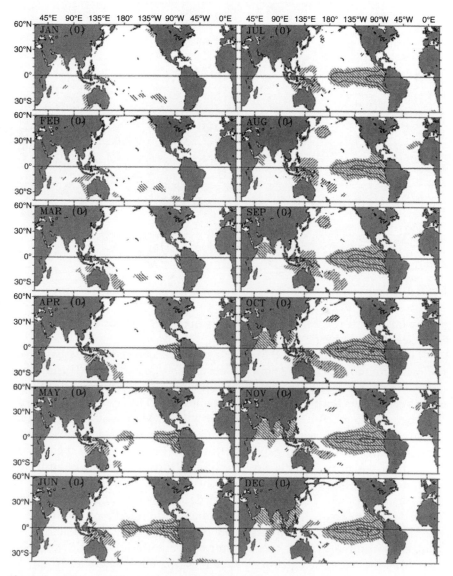

Fig. 8.3 SSTA (in °C) composites for January (0) to June (+1). The contour interval is 0.5°C. The diagonal shading northwest–southeast corresponds to a 95% significant positive anomaly while the northeast–southwest diagonal shading corresponds to a 95% significant negative anomaly. (Adapted from Harrison and Larkin (1998). D. E. Harrison and N. K. Larkin generously provided this figure.)

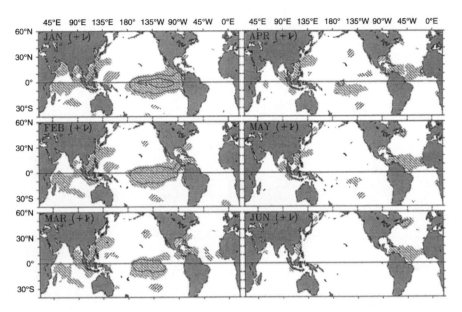

Fig. 8.3 Continued.

approximately to the El Niño episode as measured by positive NINO3.4 SST (see Fig. 8.1).

The SST anomalies (Fig. 8.3) have a similar structure to those based on the more limited data for the episodes between 1950 and 1973 (Rasmusson and Carpenter 1982). The dominant feature is the central and eastern equatorial Pacific/South American coast anomaly. This anomaly begins off the coast of South America in March (0) (i.e., March of year (0)) and spreads westward. In May (0) another positive anomaly develops in the central Pacific. In the ensuing months (June (0)–November (0)) the eastern and central equatorial Pacific positive SSTAs grow, join and intensify to reach a peak in November (0), consistent with the peak near the end of the year in Fig. 8.1. The anomaly then decays, weakening first near the South American coast. It is essentially gone by the middle of year (+1).

Although the SSTAs are largest in the eastern equatorial Pacific, we must keep in mind that the mean water temperature is much lower there than over the western equatorial Pacific warm pool where deep atmospheric convection occurs (see Fig. 2.4). Usually the eastern equatorial Pacific SSTAs are not strong enough to trigger deep equatorial convection (see the discussion in Section 6.5 and the negligible *equatorial* OLR anomalies in Figs 2.2(c) and 2.9(c)). Consequently, the zonal *equatorial* wind anomalies generated are negligible (see Fig. 8.4). Since it is the zonal equatorial winds that drive the equatorial ocean (see Section 5.2) we expect

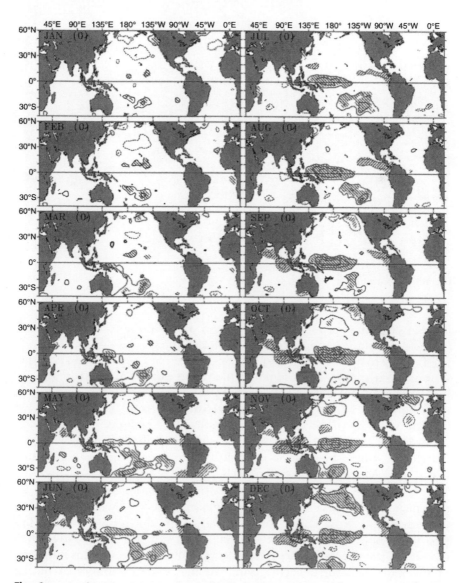

Fig. 8.4 As for Fig. 8.3 but for westerly (eastward) wind anomaly in ms⁻¹ with contour interval 0.5 m s^{-1}. (Adapted from Harrison and Larkin (1998). D. E. Harrison and N. K. Larkin generously provided this figure.)

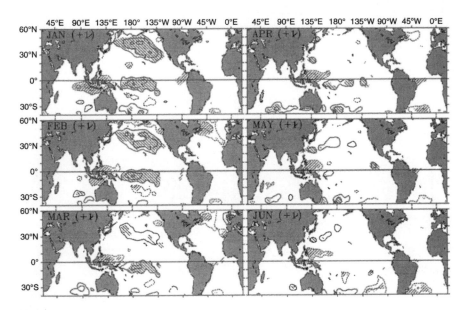

Fig. 8.4 Continued.

that any coupling between atmosphere and ocean in the eastern equatorial Pacific is of minor importance to ENSO dynamics.

Turning to other regions, some phase-locking of a negative SSTA to the seasonal cycle can be seen around northern and eastern Australia (Fig. 8.3). It appears around March (0) and in June (0) begins to intensify and grow, reaching an extremum in August (0)–October (0) and then decaying by the end of year (0). SSTAs outside the Pacific are weak. Positive SST anomalies occur in the tropical Indian Ocean from about September (0) to March (+1) and in the Northern tropical Atlantic near the location of the ITCZ from March (+1) to June (+1).

The strongest zonal wind anomalies are the anomalous westerlies in the equatorial Pacific. They are weak before June (0) but then begin to strengthen and grow and extend eastward into the central Pacific (see Fig. 8.4). They reach their peak and maximum extent in about November of year (0), then decay and are essentially gone by April (+1). This summary of the zonal equatorial Pacific wind anomalies omits important details which are not easily seen in the global perspective of Fig. 8.4. As noted earlier in Section 5.2, the ocean is extremely sensitive to low-frequency equatorial zonal wind forcing, so to understand ocean–atmosphere coupling we will study the zonal equatorial Pacific wind anomalies more closely (see Section 8.4).

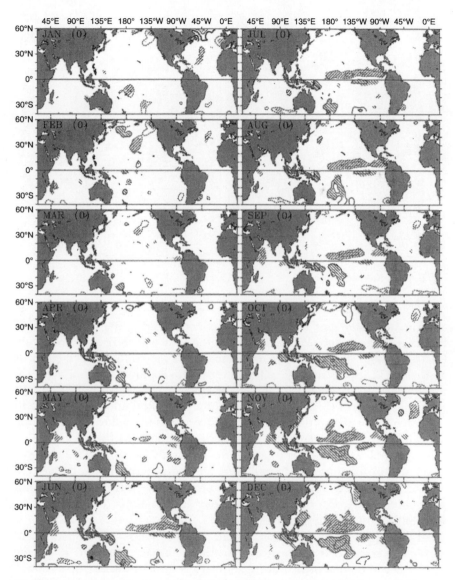

Fig. 8.5 As for Fig. 8.3 but for southerly (northward) wind anomaly in m s^{-1} with contour interval 0.5 m s^{-1}. (Adapted from Harrison and Larkin (1998). D. E. Harrison and N. K. Larkin generously provided this figure.)

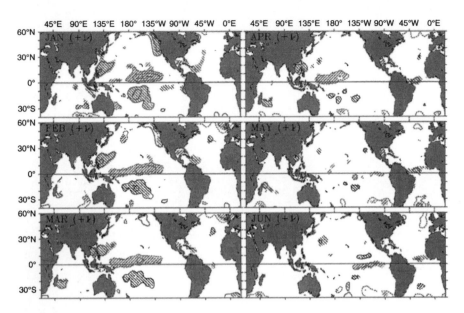

Fig. 8.5 Continued.

Another feature of the zonal Pacific wind field is the zonal band of easterlies in the eastern Pacific ITCZ region which are strongest from July (0) to October (0). These are associated with the southward movement of the ITCZ (see also Fig. 2.2). Easterly wind anomalies also occur in the Indian Ocean, especially either on or near the equator, from about September (0) to February (+1). From the dynamical arguments given in Section 6.8, we expect these easterly surface winds to be due to descending air, which probably results as the atmosphere adjusts to the air ascending in the central-western equatorial Pacific during the warm ENSO episode. Beginning in about November (0) there are also easterlies in the far western equatorial and near equatorial Pacific north of the equator. These winds peak in about April (+1) and then move further north to about 10°N, where they are still present in August (+1).

The major extratropical zonal wind anomalies are westerly and occur north and south of the equatorial heating anomaly at about 40°N and 40°S when the equatorial heating anomaly is near its peak. The strongest westerly anomalies are in the North Pacific from about December (0) to February (+1); weaker westerly South Pacific anomalies are prominent in November (0) and December (0). Westerly wind anomalies are also present in the south Pacific at around 30°S during the southern hemisphere winter months May (0)–September (0).

The meridional surface wind anomalies (see Fig. 8.5) are most prominent in the near equatorial Pacific from about June (0) to April (+1). In the eastern Pacific these winds are due to a southward movement of the ITCZ; such movement results in anomalous northerly winds at about 5°N–10°N and anomalous southerly winds within a few degrees south of the equator, especially from about June (0) to September (0). After this, from about October (0) till April (+1), the near-equatorial wind anomalies are still northerly in the northern hemisphere and southerly in the southern hemisphere, but now are more in the western and central Pacific and result from the eastward migration of the western Pacific warm pool and the deep atmospheric convection over it (see also Figs 2.2 and 2.9).

We have now looked at a broad outline of ENSO phase-locking for SSTAs and surface winds during El Niño. During La Niña, SST and zonal wind anomalies in the equatorial Pacific are opposite in sign, but outside of the Pacific and for other physical variables this simple inverse relationship does not hold. Full details are given in Larkin and Harrison (2001, 2002).

8.3. Phase-Locked ENSO Index Time Series

In the previous section we examined spatial and temporal patterns in ENSO phase-locking. Here we consider the time series structure of the key oceanic and atmospheric indices NINO3.4 and the Tahiti minus Darwin SOI.

Figure 2.14 describes the phase-locked structure of NINO3.4. The existence of the persistence barrier in April and the strong persistence for much of the rest of the year means that it may be appropriate to think of the NINO3.4 time series as being divided up into nearly independent years beginning in April and ending in March. To separate signal from noise in such an El Niño 'event' model, we will conduct an empirical orthogonal function (EOF) analysis (see Appendix A) on the monthly NINO3.4 time series. Specifically, let $X_i(t)$ be the value of NINO3.4 for calendar month i $(i = 1, 2, \ldots, 12)$ of year t $(t = 1, 2, \ldots, M)$. Form the $X_i(t)$ into a 12-element annual time series vector $\mathbf{X}(t)$ with ith element $X_i(t)$. The EOF analysis finds a 12-element 'pattern' vector $\mathbf{p}$ that maximizes the variance of the linear combination $\mathbf{p} \cdot \mathbf{X}(t)$ subject to $\mathbf{p} \cdot \mathbf{p} = 1$. More precisely, the EOF analysis finds 12 pattern vectors $\mathbf{p}^{(i)}$ that find extrema of $\sum_{t=1}^{M} (\mathbf{p} \cdot \mathbf{X}(t))^2$ subject to $\mathbf{p} \cdot \mathbf{p} = 1$; the first pattern vector $\mathbf{p}^{(1)}$ is the one that maximizes the variance. The annual time series vector $\mathbf{X}(t)$ can be written (see (A.13) and (A.10)) as

$$\mathbf{X}(t) = \sum_{i=1}^{12}(\mathbf{p}^{(i)} \cdot \mathbf{X}(t))\mathbf{p}^{(i)}. \tag{8.1}$$

In the case of NINO3.4, 86% of the variance is described by the first term in (8.1) and only 7% by the second so that we may write

$$\mathbf{X}(t) \approx (\mathbf{p}^{(1)} \cdot \mathbf{X}(t))\mathbf{p}^{(1)}. \tag{8.2}$$

In terms of components, (8.2) is of the form

$$X_i(t) \approx S_i A(t), \tag{8.3}$$

i.e., 86% of the variance of NINO3.4 can be described by an annual time series multiplied by a 12 calendar month structure function S_i which is the same every year. This implies that while it is often said that no two ENSO events are exactly alike, as far as NINO3.4 is concerned, to within an excellent approximation, the only difference is that different years have a different amplitude $A(t)$. Torrence and Webster (1998, 2000) have also considered 'event' models like (8.3) but they included an additive remainder term $R(t)$ and chose S_i differently.

Figure 8.6 shows the calendar year phase-locked EOF structure function S_i, the annual time series $A(t)$ for the years 1950–2003 and a comparison of the right-hand side of (8.3) with NINO3.4. The very good agreement emphasizes the very strong phase-locking of NINO3.4 to the seasonal cycle.

The annual amplitude time series provides a convenient way to categorize years as El Niño, La Niña or neutral. Figure 8.6 uses $A > 0.367$ for El Niño, $A < -0.367$ for La Niña and $|A| \leq 0.367$ for neutral years. Using this definition, the value of $A(t)S_i$ for the three consecutive largest S_i (November, December and January) exceeds 0.5°C for El Niño and is less than −0.5°C for La Niña. Besides categorizing past El Niños, the product $A(t)S_i$ for NINO3.4 also provides a simple way to predict whether a given year will be an El Niño year. Figure 8.7 shows that, by July, least-squares fitting $A(t)S_i$ to NINO3.4 after April gives a good estimate of $A(t)$ for that year and hence a good idea whether an El Niño will occur. Note, however, that consistent with the existence of the boreal spring persistence barrier, data before the spring are useless for predicting $A(t)$; the correlation of $A(t-1)$ with $A(t)$ is only −0.1.

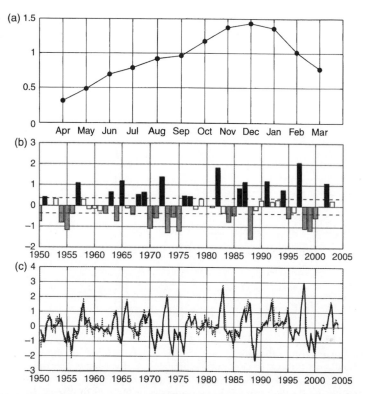

Fig. 8.6 Representation of the El Niño index NINO3.4 as the product $S_i A(t)$ where S_i ($i=1=$ January, $i=2=$ February, ..., $i=12=$ December is the calendar year structure function and $A(t)$ ($t=1950, 1951, ..., 2003$) is the annual time series amplitude function. The product $S_i A(t)$ was determined as the first mode of an EOF analysis. (a) The structure function S_i. The structure function has been normalized so that $\sum_{i=1}^{12} S_i^2 = 12$. (b) The annual amplitude function $A(t)$ with El Niño (black), La Niña (gray) and Neutral (white). The variance of $S_i A$ summed over all 12 calendar months explains 86% of the NINO3.4 variance. The horizontal dashed lines are the threshold lines $A(t) = \pm 0.367$. When $|A| > 0.367$ then $|S_i A|$ exceeds 0.5°C for at least 3 months of the year and the year is an El Niño ($A > 0$) or La Niña ($A < 0$) year. (c) NINO3.4 (dashed line) and $S_i A(t)$ (heavy solid line). The correlation coefficient between these two time series is $r = 0.93$.

A similar EOF analysis to NINO3.4 but now for the Tahiti minus Darwin SOI is shown in Fig. 8.8. The SOI is a much noisier time series and the correlation of the estimate $A(t) S_i$ with the monthly SOI time series is lower ($r = 0.74$) than NINO3.4 ($r = 0.93$). However, the correlation of the SOI $A(t) S_i$ with a 5-month running mean of the SOI is 0.89

and the annual time series of the SOI and NINO3.4 are well correlated ($r = -0.92$). The SOI phase-locking is stable over more than a century – the structure functions S_i for the first and second halves of the record are nearly identical (Fig. 8.9).

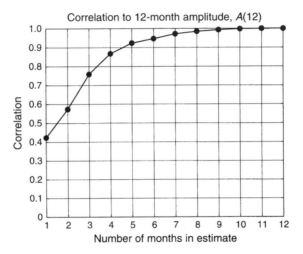

Fig. 8.7   Correlation of the amplitude A to an estimate A_* of A where A_* for April is the April value of NINO3.4 (1-month estimate), A_* for May is determined from the least-squares fit $A_* S_i$ to the April and May values of NINO3.4 (2-month estimate), A_* for June is determined from the least-squares fit $A_* S_i$ to the April, May and June values of NINO3.4 (3-month estimate), etc. By the end of July (4-month estimate) the high correlation (0.87) shows that A_* is an excellent estimate for A. For March all 12 calendar months are used in the fit and so $A_* = A$ and the correlation $= 1$.

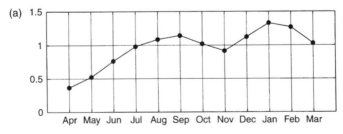

Fig. 8.8   As for Fig. 8.6 but now for the Tahiti minus Darwin SOI for the ENSO years beginning Apr 1876 and ending Mar 2004 ($t = 1876$ to $t = 2003$). (a) The structure function S_i; (b) the amplitude function $A(t)$; (c) comparison of $A(t)S_i$ with a 5-month running mean of the SOI. The SOI is a noisy time series – the variance of a 5-month running mean of the SOI is only 62% of the raw monthly time series. The product $A(t)S_i$, determined from an EOF of the raw SOI, is correlated at $r = 0.74$ with the raw SOI and $r = 0.89$ with the 5-month running mean filtered SOI.

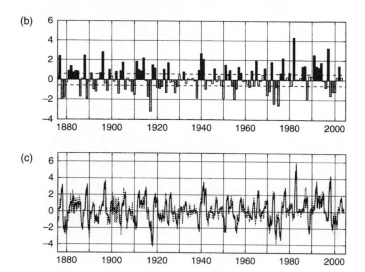

Fig. 8.8 Continued.

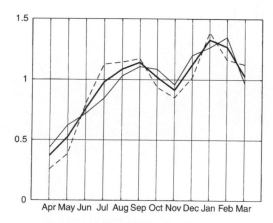

Fig. 8.9 The structure function S_i for the first half (dashed line) of the SOI record (Apr 1876–Mar 1940), the second half (light solid line) of the SOI record (Apr 1940–Mar 2004) and the whole record (heavy solid line).

8.4. PHASE-LOCKED PROPAGATING ZONAL EQUATORIAL WIND ANOMALIES

We saw in Chapter 5 that zonal equatorial wind anomalies play a key role in ENSO ocean dynamics. In this section we will examine how

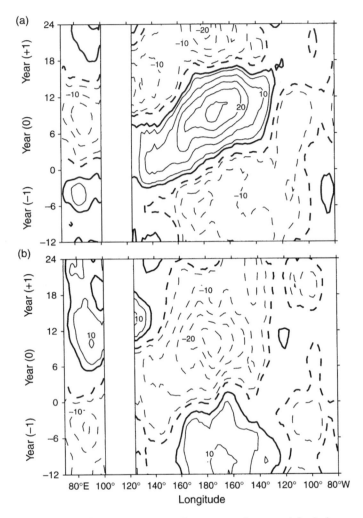

Fig. 8.10   (a) Top panels: Average sea surface eastward equatorial wind stress anomaly (in mPa) for the El Niño year (year (0)), the year before the El Niño year (year (−1)) and the year after the El Niño year (year (+1)). The average is based on the 6 El Niños 1972, 1976, 1982, 1987, 1991 and 1997. Positive contours are solid and negative contours are dashed. The heavy solid contour is 2.5 m s⁻¹, the heavy dashed contour is −2.5 m s⁻¹ and the zero contour is not plotted. The other solid line contours are 5, 10 mPa, etc., and the other dashed line contours −5, −10 mPa, etc. The gap between the panels is due to the presence of land. (b) Bottom panels. As for (a) but now for La Niña based on the seven La Niñas 1973, 1975, 1984, 1988, 1995, 1998 and 1999. (Redrawn from Clarke and Van Gorder, 2003.)

these wind anomalies are phase-locked to the seasonal cycle and then, in Section 8.5 and Section 8.6, how they are related to persistence in NINO3.4 and the SOI.

Figure 8.10(a) shows a composite of the zonal equatorial wind anomalies for the year before an El Niño year, the El Niño year and the year after an El Niño year. As we have seen earlier (see, e.g., Figs 2.2 and 2.9), El Niños are characterized by anomalous westerly winds in the west-central equatorial Pacific. Figure 8.10(a) not only shows these winds, but also indicates that they are part of a seasonally phase-locked signal that begins in the eastern Indian Ocean in the middle of the year before an El Niño and then moves eastward, reaching the western Pacific near the beginning of the El Niño year and then amplifying in the central equatorial Pacific over the last half of the year. The signal dies out over the first part of the year after the El Niño. This phase-locked propagating signal was first noticed by Barnett (1983) and was also documented by Gutzler and Harrison (1987) and Meehl (1987). During La Niña (Fig. 8.10(b)), wind anomalies of opposite sign also propagate eastward and amplify in the central equatorial Pacific.

Notice that the phase-locked propagating signal appears in the eastern equatorial Indian Ocean and western equatorial Pacific *prior to* the boreal spring, suggesting that it may be possible to overcome the persistence barrier using the wind anomaly information. Clarke and Van Gorder (2003) defined a monthly wind predictor index τ by an integration in space and time along the propagation path (see Table 8.1) to maximize the signal. This index, which is available for all calendar months (not just the El Niño and La Niña composites) is a good El Niño predictor at large leads; for 11 out of 12 months of the calendar year, it leads NINO3.4 with a correlation of 0.5 or greater for at least some lead in the range 10–15 months (see Fig. 8.11). It can be used to cross the spring persistence barrier since the correlation of January τ with the following July NINO3.4 is 0.59.

Figure 8.10 shows that, at least for the El Niño and La Niña years, the zonal wind anomalies in the central equatorial Pacific are of opposite sign to those of the eastern Indian Ocean/far western equatorial Pacific. This tendency for all years can be seen in Fig. 8.12 which shows how the index τ persists for each calendar month. The correlation suddenly changes sign in July because then the index changes its spatial location from the west-central equatorial Pacific to the eastern Indian Ocean where the zonal wind anomalies tend to be of opposite sign.

Why should the zonal equatorial wind be of opposite sign in these locations? We shall see in Chapter 9 that for dynamical reasons anomalous rising air must be compensated zonally with anomalous sinking air. Specifically, during El Niño when there is heavier than normal precipitation and anomalous heating and rising air in the central Pacific, the eastern Indian

Table 8.1 Time and space averaging of the equatorial eastward wind stress anomaly used to construct the predictor $\tau(t)$ for each calendar month. (Taken from Clarke and Van Gorder, 2003.)

Calendar month	Space and time average	
	Indian	Pacific
July	75°E–100°E, Jul	
August	75°E–100°E, Jul–Aug	
September	75°E–100°E, Jul–Sep	
October	75°E–100°E, Jul–Oct	
November	75°E–100°E, Jul–Oct	129°E–159°E, Nov
December	75°E–100°E, Aug–Oct	129°E–159°E, Nov–Dec
January	75°E–100°E, Sep–Oct	129°E–159°E, Nov–Jan
February	75°E–100°E, Oct	129°E–159°E, Nov–Feb
March		129°E–159°E, Nov–Mar
April		133°E–163°E, Dec–Apr
May		137°E–167°E, Jan–May
June		141°E–171°E, Feb–Jun

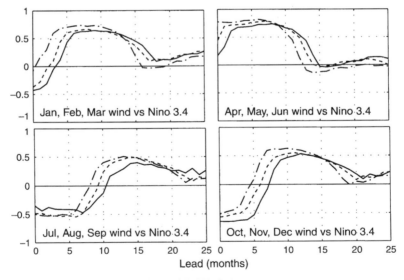

Fig. 8.11 Lead correlations of the Indo–Pacific precursor $\tau(t)$ for each calendar month with NINO3.4 (t + lead) for the period January 1971–December 2001. A positive lead corresponds to τ leading NINO3.4. The first month in each quarter is solid, the second is dashed and the third is dot-dashed. (Redrawn from Clarke and Van Gorder, 2003.)

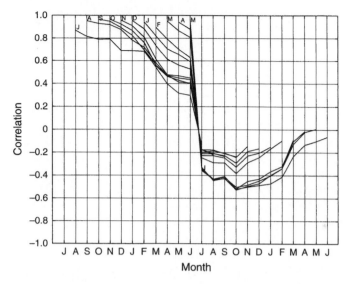

Fig. 8.12 The correlation of the Indo-Pacific wind predictor index τ for calendar month i with calendar month $i + j$ for $i = 1, 2, \ldots, 12$ and $j = 1, 2, \ldots, 12$. For example, for $i = 1$ and $i + j = 3$ the correlation is between the January τ time series and the March τ time series. In the figure each curve is labeled by the first letter of its starting month i and the correlation begins with the correlation between the i and $i + 1$ calendar month time series. For example, the first point on the curve labeled 'F' corresponds to the correlation between the February and March τ time series. The sudden drop in the correlation from June to July occurs because in July the index changes its spatial location from the west-central Pacific to the Indian Ocean where the wind anomalies tend to be of opposite sign.

Ocean/far western equatorial Pacific air tends to sink anomalously and there is less precipitation. Based on the physical arguments of Section 6.8.3, on ENSO spatial scales anomalously rising air results in westerly surface wind anomalies and anomalously sinking air results in easterly surface wind anomalies. Thus the zonal equatorial wind anomalies are of opposite sign in the two locations. Similar arguments can be given during La Niña – in that case the central equatorial surface wind anomalies are easterly as the air above sinks anomalously while in the far western equatorial Pacific the air aloft is rising anomalously and the surface equatorial winds are anomalously westerly.

But why should these zonal wind anomalies 'propagate' eastward, be phase-locked to the seasonal cycle and amplify in the central equatorial Pacific? While mechanisms have been suggested for the phase-locking of Indonesian Indian Ocean wind anomalies south of the equator to the seasonal cycle (Braak 1919; Nicholls 1978, 1979, 1984; Hendon 2003) as well as the phase-locking of equatorial Pacific wind anomalies (Clarke et al. 1998; Clarke and Shu 2000) and phase-locked propagation (Meehl 1987), currently there is no widely accepted theory for the eastward propagation seen in Fig. 8.10.

Although the propagation is not understood, perhaps the amplification of the anomalies in the central Pacific is explained by the coupled ocean–atmosphere growth mechanism of Gill and Rasmusson (1983) (see also Section 7.2.2). Specifically, anomalous westerly equatorial winds will generate anomalous eastward equatorial currents, resulting in an eastward displacement of the warm pool. The deep atmospheric convection will move with the warm pool so now there is anomalous atmospheric heating east of the normal warm pool edge. By the mechanism of Section 6.8.3, anomalous heating results in anomalous westerly equatorial winds, leading to further eastward warm pool displacement, etc. An amplification of the anomaly therefore occurs in the central equatorial Pacific near the edge of the western equatorial Pacific warm pool. This mechanism is described mathematically by the exponential growth found using (7.2) with the right-hand side given by (7.3).

A similar amplification mechanism may explain the intensification of the equatorial easterly wind anomalies near the International Date Line during La Niña (see Fig. 8.10(b)). In that case anomalous easterly equatorial winds will generate anomalous westward equatorial currents, resulting in a westward displacement of the equatorial warm pool. At a given west-central equatorial Pacific location, the SST falls as the warm pool moves westward and precipitation decreases. By the mechanism of Section 6.8.3, anomalous cooling results in anomalous easterly equatorial winds, leading to further westward warm pool displacement and an amplifying anomaly.

In summary, we have at best only a partial explanation of the seasonally phase-locked, eastward propagating equatorial zonal wind anomalies that amplify in the central equatorial Pacific. In the next section we shall see that the phase-locked structure is closely linked to the two major ENSO indices NINO3.4 and the SOI.

8.5. LINKING THE PACIFIC EQUATORIAL WIND STRESS ANOMALIES WITH THE SOI AND NINO3.4

In this section we will show that the SOI and NINO3.4 are both proportional to an integral of the zonal equatorial wind stress anomaly from the western to the eastern boundary of the Pacific. Thus if we understood the phase-locked equatorial zonal wind stress anomaly we would understand the phase-locked properties of the major ENSO indices.

8.5.1. The relationship between the Southern Oscillation and zonal equatorial wind stress anomalies

Examination of Figs 2.2(a) and 2.9(a) shows that anomalous ENSO atmospheric pressure at Darwin (12°20′S, 130°50′E) and Tahiti (17°45′S, 149°30′W) is representative, respectively, of anomalous ENSO atmospheric pressure in the western and eastern equatorial Pacific. Therefore an Equatorial Southern Oscillation Index (ESOI), the difference in anomalous atmospheric pressure between the eastern and western equatorial Pacific, will have essentially the same properties as the traditional Tahiti minus Darwin SOI. We will focus on the ESOI because, as we shall see, it is strongly linked dynamically to the zonal equatorial wind stress. The analysis below follows that of Clarke and Lebedev (1996).

Consider the zonal momentum equation in the atmosphere at the equator. Since the Coriolis parameter vanishes there, this equation may be written as

$$\rho \frac{\partial u}{\partial t} + \rho \mathbf{u} \cdot \nabla u = -p_x + X_z. \tag{8.4}$$

Here $\mathbf{u} = (u, v, w)$ is the wind velocity vector in the x (eastward), y (northward) and z (upward) directions, ∇ the three-dimensional gradient operator, ρ the air density, p the air pressure due to the motion and X the eastward turbulent stress above the water. ENSO frequencies are low enough that the $\rho(\partial u / \partial t)$ term in (8.4) is negligible. The analysis by Clarke and Lebedev (1996) also shows that at ENSO and decadal frequencies the primary balance in (8.4) is

$$p_x = X_z \tag{8.5}$$

with p_x nearly independent of z over the surface stress boundary layer. Integrating this equation from the surface (where $z = 0$ and $X = \tau^x$, the wind stress) to the boundary layer height d where X vanishes, we get

$$-\tau^x \approx dp_x. \tag{8.6}$$

Since d is only weakly dependent on x, integration of (8.6) from the western ($x = 0$) to the eastern ($x = L$) Pacific gives

$$-\int_0^L \tau^x \, \mathrm{d}x = d\Delta p \tag{8.7}$$

where Δp, the difference in surface atmospheric pressure between the eastern and western Pacific, is the ESOI. A linear regression of interannual $\Delta p(t)$ with interannual $\int_0^L \tau^x \, dx$ (see (8.7)) gives a correlation coefficient $r = -0.9$ and a regression coefficient with boundary layer height $d \approx 680 \, m$ (Clarke and Lebedev 1997). Thus the ESOI and zonal wind stress anomaly are linked through (8.7).

8.5.2. The relationship between El Niño and the zonal equatorial wind stress

Observations (Yu and McPhaden 1999) suggest that at ENSO frequencies at the equator and near the ocean surface the zonal momentum equation in the ocean has the same form as (8.5), viz.,

$$p_x = X_z \tag{8.8}$$

where now X is the eastward anomalous turbulent stress *in* the water near the surface and p is the anomalous ocean pressure due to the anomalous stress. Near the surface, by hydrostatic balance, this pressure is due to extra weight associated with anomalous sea level and so

$$p = \rho_* g \eta \tag{8.9}$$

with $\rho_* = $ mean water density. Thus if we integrate (8.8) over the mixed layer of depth H_{mix} we have, using (8.9),

$$H_{\text{mix}} \rho_* g \eta_x = \tau^x \tag{8.10}$$

where we have used $X = \tau^x$ at the sea surface and $X = 0$ at the bottom of the ocean mixed layer.

Eastward integration of (8.10) from the western Pacific boundary $(x = 0)$ to the eastern Pacific boundary $(x = L)$ gives

$$\rho_* g H_{\text{mix}} [\eta(L) - \eta(0)] = \int_0^L \tau^x \, dx. \tag{8.11}$$

This balance can be generalized (Li and Clarke 1994) to include time delays associated with ocean waves adjusting the ocean to the quasi-equilibrium balance (8.8). Both (8.11) and the analogous thermocline slope result (5.50) have been verified observationally for ENSO time scales (Kessler 1990; Li and Clarke 1994).

Since the equatorial ocean behaves, approximately, as a two–layer fluid with the thermocline as an interface between the two layers, the sea level difference $\eta(L) - \eta(0)$ is proportional to the thermocline slope across the Pacific. In fact the thermocline is of finite thickness and its upper part affects SST in the eastern equatorial Pacific (see Fig. 2.7). Thus interannual changes in the thermocline slope result in interannual changes in SST in the eastern equatorial Pacific and hence NINO3.4. This can be seen in Figs 5.3 and 5.4 which show that the slope of the sea level and thermocline vary like NINO3.4. Since the thermocline and sea level slopes vary interannually like NINO3.4, by (8.11) NINO3.4 varies like $\int_0^L \tau^x \, dx$.

8.5.3. Phase-locked structure of NINO3.4 and the SOI

Since NINO3.4 is proportional to $\int_0^L \tau^x \, dx$, we can see how its phase-locked structure arises graphically for El Niño by integrating the composite wind stress in Fig. 8.10(a) from the western equatorial Pacific (starting just after the gap in the figure) to the eastern equatorial Pacific along lines parallel to the abscissa. We see that the increasing NINO3.4 amplitude in year 0 from April to December and then the decrease to March the following year (see Fig. 8.6(a)) is largely due to the growing and then decaying zonal equatorial wind anomaly in the central equatorial Pacific. Similar results apply to La Niña (Fig. 8.10(b)); in this case the winds are of opposite sign and the amplitude decrease to March is not as steep.

Similar results also apply for the SOI. Thus the growth in the amplitude of the structure functions from April to the end of the year and also the strong persistence from June onward (see Fig. 2.14) are related to the growth in wind anomaly amplitude in the central equatorial Pacific. A coupled ocean–atmosphere mechanism for this anomaly growth has been proposed (see Section 8.4) but this mechanism by itself does not explain the phase-locking to the seasonal cycle.

▶ 8.6. PHASE-LOCKING AND BIENNIAL VARIABILITY

Suppose that the phase-locking of the zonally propagating equatorial wind stress anomalies were perfect. Then an anomaly of a given sign would begin in the Indian Ocean in July, 'propagate' eastward into the central Pacific by June of the following year and generate an anomaly of opposite sign in July in the eastern Indian Ocean. Such a model, having succeeding year-long propagating winds of opposite sign, exhibits biennial (2-year) variability. Many researchers have discussed a 'tropospheric biennial oscillation' and its linkage to El Niño (see, e.g., Rasmusson et al. (1990) for a summary of the early work and a comprehensive analysis of the observations). Brier (1978), Nicholls (1978, 1979, 1984), Meehl (1987, 1993, 1994, 1997), Clarke et al. (1998), Clarke and Shu (2000) and Hendon (2003) have suggested possible tropospheric biennial mechanisms and Clarke and Van Gorder (1999) and Weiss and Weiss (1999) have linked tropospheric quasi-biennial variability to the boreal spring persistence barrier.

But in fact the phase-locking of the zonally propagating equatorial wind stress anomalies to the seasonal cycle is far from perfect and so the biennial variability is very weak. This is shown in Fig. 8.12. If the wind index were purely biennial, then for each calendar month the correlation should be −1 at a lag of 1 year. Therefore in Fig. 8.12 the 12 plots, each 1 year long, should end with correlation at −1 but instead they end at about −0.2 or closer to zero.

Because (see Section 8.5) NINO3.4 and the SOI are approximately proportional to $\int_0^L \tau^x dx$, we expect that, like the eastward equatorial wind anomaly τ^x, the NINO3.4 and SOI indices will also be only very weakly biennial. If NINO3.4 were purely biennial, then the annual function $A(t)$ in the NINO3.4 signal $A(t)S_i$ (see Fig. 8.6) should be correlated with $A(t+1)$ with correlation coefficient −1 but in fact the correlation is nearly zero ($r = -0.11$). The corresponding correlation for the SOI annual time series is also nearly zero ($r = -0.02$). Thus ENSO is more like a sequence of statistically independent events than a biennial oscillation. Even though this is the case, we should not think that ENSO event years are random and cannot be anticipated; the Indo–Pacific wind predictor in Fig. 8.11 and the warm water volume in the equatorial Pacific (see Section 11.3) both foreshadow NINO3.4 and the SOI before April, the beginning of the ENSO year for these indices.

8.7. A Phase-Locking Mechanism for the Termination of ENSO Events

As we have seen earlier in this chapter, there is a tendency for the ENSO signal to peak near the end of the calendar year (see Figs 8.1, 8.2 and 8.6). Why should this be so? In this section we discuss why ENSO events should tend to stop growing and begin to decrease near the end of the calendar year.

8.7.1. Basic mechanism

The panels July (0) to March (+1) of Fig. 8.4 show that typically during El Niño the westerly wind anomalies in the west-central equatorial Pacific are centered on the equator during July to December and then move southward of the equator in January to March. This southward shift can also be seen on the front cover of this book for the 1997 El Niño. The southward shift of the equatorial wind anomalies during El Niño was first noticed by Harrison (1987). Harrison and Vecchi (1999) pointed out that this southward shift results in 'greatly reduced equatorial wind anomalies' which, they showed using a numerical model, directly resulted in a shoaling thermocline in the eastern equatorial Pacific. This wind-forced shoaling is consistent with the dynamics discussed earlier in Section 8.5.2 and also in Sections 5.4.2 and 5.5.

The shoaling thermocline in the eastern equatorial Pacific implies decreased SST and the calendar year phase-locked fall in NINO3.4 in January, February and March seen in Figs 8.1 and 8.6. The southward movement of easterly equatorial wind anomalies also occurs during La Niña at the same time of the calendar year (Larkin and Harrison 2002) so a similar explanation of opposite sign applies for the La Niña phase-locked demise seen in January–March in Fig. 8.2.

Based on the above, the key to understanding a major component of the mysterious ENSO phase-locking to the calendar year, namely, the demise of NINO3.4 in January, February and March, is to understand why the equatorial winds shift southward of the equator. Harrison and Vecchi (1999) pointed out that the western equatorial Pacific cools from October/November to December/January as the overhead position of the sun moves southward. They suggested that the deep atmospheric convection and westerly wind anomalies follow the warmest water and that in this way, near the end of the calendar year, the westerly wind anomalies move southward of the equator. (This southward warm water movement and associated southward shift of the westerly wind anomalies is illustrated on the front cover of this book for the 1997–1998 El Niño). As the westerly winds on the equator decrease markedly, the anomalous downward thermocline tilt toward the east decreases, the eastern equatorial Pacific thermocline shoals and the peak of El Niño is thus phase-locked to the calendar year. Note that the phase-locking depends critically on the total SST and therefore on the southward seasonal march of the SST. Total SST rather than just the SST anomaly is key to deep convection based on the arguments given in Chapter 6.

8.7.2. Large El Niños

Lengaigne et al. (2006) examined the behavior of the monthly NINO3 SST index (the average monthly SST anomaly in the eastern equatorial Pacific region 5°S–5°N, 150°W–90°W) for the period 1871–2002 and found that strong El Niños, i.e., those El Niños having a mean fall SST anomaly greater than 1.25°C, behaved differently to moderate El Niños (mean fall anomaly greater than 0.5°C and less than 1.25°C). Specifically, strong El Niños peak in boreal winter, decay rapidly in boreal spring and summer and tend to be followed by La Niña conditions. Moderate El Niños, on the other hand, while often exhibiting a local maximum at the end of the calendar year, are followed by La Niña, neutral or El Niño conditions. La Niñas, strong or moderate, behave similarly to moderate El Niños.

Why should strong El Niños behave differently to moderate El Niños and La Niñas? When very strong El Niños occur, the equatorial warm pool penetrates far into the eastern equatorial Pacific and carries with it anomalous deep equatorial atmospheric convection and equatorial westerly wind anomalies (see Fig. 6.1). These westerly wind anomalies reduce the speed of the easterly equatorial trade winds and the associated upwelling and vertical mixing. In their analysis of the huge 1997 El Niño, Vecchi and Harrison (2006) and Vecchi (2006) pointed out that the southward shift of the central equatorial westerlies at the end of the year did result in a raised eastern equatorial Pacific thermocline, but because the westerly wind anomalies in the eastern equatorial Pacific had reduced the equatorial trade wind speed to nearly zero, there was negligible vertical mixing and the raised thermocline did not affect SST. However, as the northern hemisphere summer approached and the overhead sun moved northward, the reduced insolation lowered the total SST in the eastern equatorial Pacific. As a result in May the deep atmospheric convection ceased and the easterly trade winds abruptly returned and mixed the surface water and the upper part of the raised thermocline. Consequently the eastern equatorial Pacific SST in May 1998 fell by more than 4°C within 2 weeks. Vecchi (2006) noted that this mechanism was also operating in the huge 1982–1983 El Niño and that it should be expected for extreme El Niño events.

The above Vecchi and Harrison mechanism gives the very strong El Niños a strong extra kick toward La Niña in May/June in addition to the end of the year southward shift of the central equatorial Pacific zonal wind anomalies and other negative feedbacks discussed in Chapter 7. Only very strong El Niños feel this May/June kick because only they penetrate far into the eastern equatorial Pacific where the dynamics is strongly influenced by the seasonal migration of the ITCZ. This extra kick toward La Niña may thus explain the Lengaigne et al. (2006) result that strong El Niños are more likely to give rise to an event of opposite phase next year than either moderate El Niños or La Niñas.

8.8. ELEMENTS OF A SEASONALLY PHASE-LOCKED ENSO MECHANISM

Suppose, for the sake of argument, that we are near the end of the calendar year of a moderate El Niño year. By the argument of Section 8.7, the southward march of SST due to the southward march of the maximum insolation will result in a lowering of the SST on the equator and a southward displacement off the equator of the easterly wind anomalies in the central equatorial Pacific. Consequently the thermocline will rise in the eastern Pacific, surface equatorial water will cool, causing a decrease in the El Niño index and, consequently, a maximum near the end of the calendar year as observed (see Figs 8.1, 8.2 and 8.6).

What happens next depends on (i) the influence of the equatorial heat content, which is strongly correlated with ENSO across the spring (see the physical arguments of Section 7.3 and Fig. 11.3); (ii) the negative feedbacks discussed in Section 7.2; and (iii) random variability not resolved by the large-scale, low-frequency deterministic dynamics in (i) and (ii). As pointed out by Spencer (2004), because the equatorial Pacific is seasonally warmed in April, deep atmospheric convection is more likely then and the equatorial Pacific is most sensitive to SST anomalies. Via the coupled atmosphere ocean instability discussed in Sections 7.2.2 and 8.4 above, the SST anomaly, which may be positive or negative, will tend to grow. As the coupled ocean–atmosphere instability strengthens, the amplitude of NINO3.4 increases (see Fig. 8.6) until the end of the year when the seasonal cycle of insolation in the central Pacific causes the seasonally phase-locked amplitude maximum near the end of the calendar year. In this way NINO3.4 can be thought of, approximately, as always having a calendar year growth and decay structure of the form S_i multiplied by an amplitude A whose sign and size varies from year to year (see Fig. 8.6).

As we saw in Section 8.7.2, if the ENSO event is a very large El Niño, then the above mechanism is modified slightly. In that case, anomalous deep atmospheric convection occurs even in the eastern equatorial Pacific and there are two calendar year phase-locked kicks toward La Niña: the decrease in the central Pacific westerly wind anomalies at the end of the calendar year and the termination of the eastern equatorial westerly wind anomalies in May/June.

 8.9. Concluding Remarks

In this chapter we have examined some features of the phase-locking of ENSO to the seasonal cycle. Some progress has been made on describing the observations and understanding key features of the ENSO phase-locking, but our understanding is still rudimentary.

 References

Barnett, T. P., 1983: Interaction of the monsoon and Pacific trade wind system at interannual time scales. Part I: The equatorial zone. *Mon. Weather. Rev.*, **111**, 756–773.

Braak, C., 1919: Atmospheric variations of short and long duration in the Malay Archipelago and neighboring regions, and the possibility to forecast them. *Koninkljik Magnetisch en Meteorologisch.*, Verhandelingen No. 5, Observatorium te Batavia, Indonesia, 57 pp.

Brier, G. W., 1978: The quasi-biennial oscillation and feedback processes in the atmosphere–ocean–earth system. *Mon. Weather Rev.*, **106**, 938–946.

Clarke, A. J., and A. Lebedev, 1996: Long-term changes in the equatorial Pacific trade winds. *J. Clim.*, **9**, 1020–1029.

Clarke, A. J., and A. Lebedev, 1997: Interannual and decadal changes in equatorial windstress in the Atlantic, Indian and Pacific Oceans and the eastern ocean coastal response. *J. Clim.*, **10**, 1722–1729.

Clarke, A. J., X. Liu, and S. Van Gorder, 1998: On the dynamics of the biennial oscillation in the equatorial Indian and far western Pacific oceans. *J. Clim.*, **11**(5), 987–1001.

Clarke, A. J., and L. Shu, 2000: Biennial winds in the far western equatorial Pacific phase-locking El Niño to the seasonal cycle. *Geophys. Res. Lett.*, **27**(6), 771–774.

Clarke, A. J., and S. Van Gorder, 1999: On the connection between the boreal spring southern oscillation persistence barrier and the tropospheric biennial oscillation. *J. Clim.*, **12**(2), 610–620.

Clarke, A. J., and S. Van Gorder, 2003: Improving El Niño prediction using a space-time integration of Indo-Pacific winds and equatorial Pacific upper ocean heat content. *Geophys. Res. Lett.*, **30**(7), doi:10.1029/2002GL016673, 10 April 2003.

Gill, A. E., and E. M. Rasmusson, 1983: The 1982–83 climate anomaly in the equatorial Pacific. *Nature* (London), **306**, 229–234.

Gutzler, D. S., and D. E. Harrison, 1987: The structure and evolution of seasonal wind anomalies over the near-equatorial eastern Indian and western Pacific oceans. *Mon. Weather Rev.*, **115**, 169–192.

Harrison, D. E., 1987: Monthly mean island surface winds in the central tropical Pacific and El Niño events. *Mon. Weather Rev.*, **115**, 3133–3145.

Harrison, D. E., and N. K. Larkin, 1998: El Niño–Southern Oscillation sea surface temperature and wind anomalies, 1946–1993. *Rev. Geophys.*, **36**, 353–400.

Harrison, D. E., and G. A. Vecchi, 1999: On the termination of El Niño. *Geophys. Res. Lett.*, **26**, 1593–1596.

Hendon, H. H., 2003: Indonesian rainfall variability: Impacts of ENSO and local air–sea interaction. *J. Clim.*, **16**, 1775–1790.

Kessler, W. S., 1990: Observations of long Rossby waves in the northern tropical Pacific. *J. Geophys. Res.*, **95**, 5183–5217.

Larkin, N. K., and D. E. Harrison, 2001: Tropical Pacific ENSO cold events, 1946–1995: SST, SLP, and surface wind composite anomalies. *J. Clim.*, **19**, 3904–3931.

Larkin, N. K., and D. E. Harrison, 2002: ENSO warm (El Niño) and cold (La Niña) event life cycles: Ocean surface anomaly patterns, their symmetries, asymmetries, and implications. *J. Clim.*, **15**, 1118–1139.

Lengaigne, M., J.-P. Boulanger, C. Menkes, and H. Spencer, 2006: Influence of the seasonal cycle on the termination of El Niño events in a coupled general circulation model. *J. Clim.*, **19**, 1850–1868.

Li, B., and A. J. Clarke, 1994: An examination of some ENSO mechanisms using interannual sea level at the eastern and western equatorial boundaries and the zonally averaged equatorial wind. *J. Phys. Oceanogr.*, **24**, 681–690.

Meehl, G. A., 1987: The annual cycle and interannual variability in the tropical Pacific and Indian Ocean regions. *Mon. Weather Rev.*, **115**, 27–50.

Meehl, G. A., 1993: A coupled air–sea biennial mechanism in the tropical Indian and Pacific regions. Role of the ocean. *J. Clim.*, **6**, 31–41.

Meehl, G. A., 1994: Coupled land–ocean–atmosphere processes and south Asian monsoon variability. *Science*, **266**, 263–267.

Meehl, G. A., 1997: The south Asian monsoon and the tropospheric biennial oscillation (TBO). *J. Clim.*, **10**, 1921–1943.

Nicholls, N., 1978: Air–sea interaction and the quasi-biennial oscillation. *Mon. Weather Rev.*, **106**, 1505–1508.

Nicholls, N., 1979: A simple air–sea interaction model. *Q. J. R. Meteorol. Soc.*, **105**, 93–105.

Nicholls, N., 1984: The Southern Oscillation and Indonesian sea surface temperature. *Mon. Weather Rev.*, **112**, 424–432.

Rasmusson, E. M., and T. H. Carpenter, 1982: Variations in sea surface temperature and surface wind fields associated with the Southern Oscillation/El Niño. *Mon. Weather Rev.*, **110**, 354–384.

Rasmusson, E. M., X. Wang, and C. F. Ropelewski, 1990: The biennial component of ENSO variability. *J. Mar. Syst.*, **1**, 71–96.

Spencer, H., 2004: Role of the atmosphere in seasonal phase locking of El Niño. *Geophys. Res. Lett.*, **31**, L24104, doi:10.1029/2004GL021619.

Torrence, C., and P. J. Webster, 1998: The annual cycle of persistence in the El Niño/Southern Oscillation. *Q. J. R. Meteorol. Soc.*, **124**, 1985–2004.

Torrence, C., and P. J. Webster, 2000: Comments on "The connection between the boreal spring southern oscillation persistence barrier and biennial variability." *J. Clim.*, **13**, 665–667.

Vecchi, G. A., 2006: The termination of the 1997–98 El Niño. Part II: Mechanisms of atmospheric change. *J. Clim.*, **19**, 2647–2664.

Vecchi, G. A., and D. E. Harrison, 2006: The termination of the 1997–98 El Niño. Part I: Mechanisms of oceanic change. *J. Clim.*, **19**, 2633–2646.

Weiss, J. P., and J. B. Weiss, 1999: Quantifying persistence in ENSO. *J. Atmos. Sci.*, **56**, 2737–2760.

Yu, X., and M. J. McPhaden, 1999: Dynamical analysis of seasonal and interannual variability in the equatorial Pacific. *J. Phys. Oceanogr.*, **29**, 2350–2369.

CHAPTER NINE

Upper Air Response to ENSO Heating and Atmospheric Teleconnections

Contents

9.1. Overview

Although ENSO generation primarily occurs in the west-central Pacific, it causes short–term climate fluctuations in regions far from there. In this chapter we will study these 'teleconnections' to remote regions.

We begin, in Section 9.2, with a broad overview of the observed teleconnections. One of the key ENSO teleconnections is tropical rainfall which is strongly zonally asymmetric; the excess (deficit) rainfall in the central equatorial Pacific during a warm (cold) ENSO event, for example, is balanced by a rainfall deficit (excess) in the western equatorial Pacific. Section 9.3 discusses a physical reason for this zonal asymmetry. Even though the anomalous heating driving the atmosphere is nearly zonally asymmetric, much of the atmospheric response is zonally symmetric. Sections 9.4 and 9.5 describe how this can be understood physically in terms of atmospheric equatorial Kelvin and Rossby waves. ENSO heating affects the extra tropics as well as the tropics and the large, zonally symmetric component of this teleconnection is described and analyzed in Section 9.6. Section 9.7 describes and discusses theory for the zonally asymmetric component of the extratropical response. This teleconnection component has been studied extensively and we will focus on the major mechanism.

▶ 9.2. Observations

Figure 9.1, from Halpert and Ropelewski (1992), shows the typical surface air temperature teleconnections during a cold (Fig. 9.1(a)) and a warm (Fig. 9.1(b)) ENSO event. The teleconnections depend on the season and in many, but not all, cases reverse sign when the ENSO event changes sign. Over most of the tropics surface air temperature tends to be lower than normal during a cold ENSO event and higher than normal during a warm ENSO event. This suggests that the tropical air temperature response to ENSO heating anomalies has a large zonally symmetric component and Fig. 9.2(a) shows that this is indeed the case. This is in marked contrast to the tropical rainfall teleconnections, which are shown in Fig. 9.2(b) to have a strong zonally asymmetric component; the main rainfall anomalies in the central equatorial Pacific ENSO generation region are balanced zonally mainly by the rainfall anomalies in the western equatorial Pacific.

The zonally symmetric component of the temperature anomaly is of one sign throughout the tropical troposphere (Fig. 9.3). From (6.51) and (6.52) it follows that the thickness ΔZ of the atmospheric layer between (say) a near-surface pressure of 1000 hPa and an upper air 200 hPa pressure surface should increase during a warm ENSO event and decrease during a cold ENSO event. Since ΔZ is dominated by the upper air height, the 200 hPa height anomaly should have a substantial zonally symmetric component which goes up during a warm ENSO event and down during a cold ENSO event. Horel and Wallace (1981) recognized this and used long records of tropical 200 hPa data from 10 stations around the world and a summary of previous work to show that this is true. Figure 9.4 (from DeWeaver and Nigam 2002) also indicates this and, in addition, shows that poleward of the tropical height anomalies are extratropical anomalies of opposite sign with a substantial zonally symmetric component. This suggests that when (say) a warm ENSO event occurs, much of the tropical troposphere warms and much of the mid-latitudes cool. The associated geopotential 200 hPa heights imply, by geostrophy, that at 200 hPa there

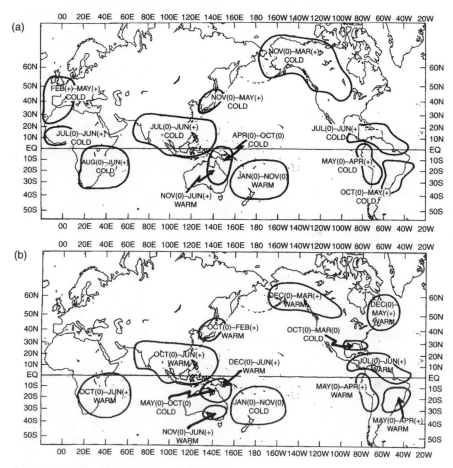

Fig. 9.1 Surface air temperature patterns during a (a) cold ENSO event and (b) warm ENSO event. The analysis is based on temperature data from about 1200 land stations. (From Halpert and Ropelewski 1992.)

are westerly wind anomalies in approximately the 20–30° latitude band. Opposite anomalies tend to occur during cold ENSO events.

The above discussion has emphasized the symmetric component of the anomalous height and air temperature ENSO response. But Fig. 9.4 also shows that there is considerable zonal asymmetry; the 200 hPa height anomaly is much larger during the Northern Hemisphere winter at about 130°W than in the western Pacific. The 200 hPa height is in the form of a 'dumbbell' shape at this longitude with peaks roughly symmetric about the equator at 15°S and 15°N. As one might expect, similar peaks are observed in the tropospheric temperature anomaly (Yulaeva and Wallace

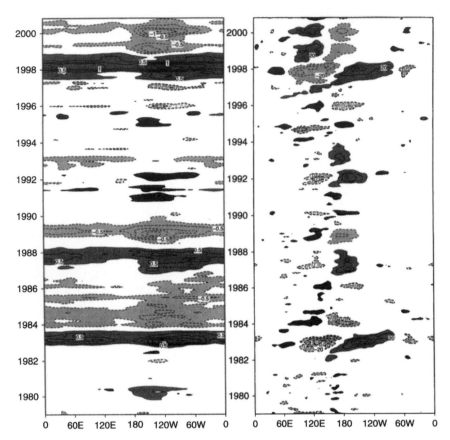

Fig. 9.2 (Left) Longitude time plots of microwave sounding unit channel-2 (MSU-2) temperature anomalies and (right) anomalies of negative out-going long-wave radiation (OLR). The MSU-2 data are a proxy for temperature anomalies averaged over the depth of the troposphere, while negative OLR is a proxy for deep convective latent heating. The positive (negative) values in the right panel represent anomalous heating (cooling) of the atmosphere via deep atmospheric convection. The MSU-2 and OLR data are based on the latitude belt extending from 10°N to 10°S. Contour intervals are 0.25 K and 10 W m^{-2}, respectively. Dark shading represents positive values and light shading negative values. (Adapted from Clarke and Kim 2005; see also Yulaeva and Wallace 1994.)

1994). Even though the dumbbell shape is formed by anomalies, i.e., the seasonal cycle has been removed, Yulaeva and Wallace (1994) showed that the amplitude of the dumbbell shape is strongly phase locked to the seasonal cycle, being much stronger in February–March–April than in August–September–October.

The polar projection plots of Fig. 9.5 show 200 mb height estimates of ENSO teleconnections poleward of 20°N based on eight warm and

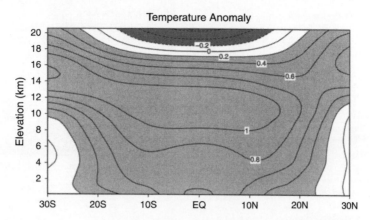

Fig. 9.3 Vertical structure of the zonally averaged interannual temperature anomaly found by regressing the monthly time series of anomalous monthly temperature T' onto the NINO3.4 index and then zonally averaging. The NINO3.4 index was normalized by $\sqrt{2}$ times its standard deviation. The contours are in °C and values greater than 0.2°C or less than −0.2°C are shaded. (Figure provided by Dr. K.-Y. Kim.)

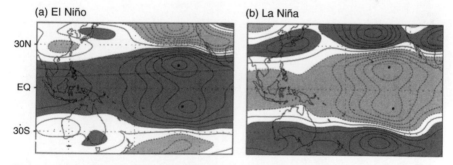

Fig. 9.4 The 200 hPa height anomalies for (a) eight El Niño winters and (b) eight La Niña winters. 'Winter' is defined as the four-month period December, January, February and March. The eight El Niño winters began in December for the years 1957, 1965, 1968, 1972, 1982, 1986, 1991 and 1997 while seven of the eight La Niña winters began in December for the years 1970, 1973, 1975, 1984, 1988, 1998 and 1999. The remaining winter began in December 1949 but since the data began in January 1950 that winter was based on data for January–March 1950. (c) Difference of the composites in (a) and (b), used to represent the 'linear' component of the ENSO response, which changes only in sign between the warm and cold composites. (d) Sum of (a) and (b), used to represent the "non-linear" component of the response, illustrating differences between warm and cold event responses. The solid dots represent the centers of the North Pacific and South Pacific highs. The contour interval is 10 m, with dark (light) shading for positive (negative) values in excess of 10 m. (From DeWeaver and Nigam 2002.)

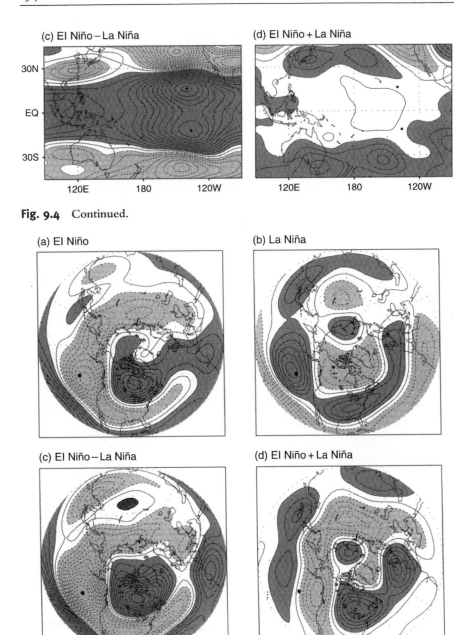

Fig. 9.4 Continued.

Fig. 9.5 As for Fig. 9.4 but now for the Northern Hemisphere, beginning at 20°N. The solid dot represents the center of the northeast Pacific low. (From DeWeaver and Nigam 2002.)

eight cold events. If El Niño and La Niña were exactly opposite in phase with the same structure and amplitude, then Fig. 9.5(c) would be exactly twice Fig. 9.5(a) and Fig. 9.5(d) would be zero. This is clearly not the case, but the difference in Fig. 9.5(c) generally dominates over the sum in Fig. 9.5(d) showing that the warm and cold event structures tend to have a similar amplitude and opposite phase. This is particularly the case near the tropics. The northeast Pacific anomaly also has nearly the same structure and amplitude and opposite sign for warm and cold events; there is only a slight (about 10° of longitude) westward phase shift during a La Niña winter. On the other hand, while the anomaly centered over northern North America is of opposite sign for warm and cold events, the warm event anomaly is centered over southern Hudson's Bay with a peak amplitude of about 70 m and the cold event anomaly has an amplitude of only about 30 m and is shifted northward.

Note that the above broad–brush teleconnection estimates are *averages* based on eight El Niños and eight La Niñas. When (say) an El Niño occurs, some of the teleconnections, particularly those in the extra tropics, may not be observed. This may be due to the presence of another climate signal, like the North Atlantic Oscillation, or irregular random variability.

9.3. NEARLY ZONALLY ASYMMETRIC ENSO HEATING ANOMALIES

In order to learn more about ENSO teleconnections, we will begin in the tropics where the signal is strongest and focus first on the structure of the ENSO heating anomalies which drive the flow. Probably the most striking aspect of this structure is that it nearly balances zonally (see Fig. 9.2, right panel). Why should this be so?

Following Clarke and Kim (2005), consider any constant equatorial pressure surface S that circles the globe between (say) 10°S and 10°N. Integrate the continuity equation (6.64) over the surface S and use Gauss's divergence theorem to get

$$\int_{10°N} v\, dx - \int_{10°S} v\, dx + \int_S \frac{\partial \omega}{\partial p}\, dS = 0. \tag{9.1}$$

If v were exactly geostrophic at 10° latitude then the first two integrals on the left-hand side of (9.1) would vanish. Therefore, to within a small error of order the Rossby number at 10°N and 10°S, (9.1) can be written as

$$\int_S \frac{\partial \omega}{\partial p}\,\mathrm{d}S = 0. \tag{9.2}$$

Since S is a constant pressure surface, it follows from (9.2) that

$$\frac{\partial}{\partial p}\left(\int_S \omega\,\mathrm{d}S\right) = \int_S \frac{\partial \omega}{\partial p}\,\mathrm{d}S = 0. \tag{9.3}$$

Thus $\int_S \omega\,\mathrm{d}S$ is independent of pressure (and therefore height). But at the earth's surface the vertical velocity is zero and consequently, by (6.33) and (6.63), ω and hence $\int_S \omega\,\mathrm{d}S$ are zero. Therefore

$$\int_S \omega\,\mathrm{d}S = 0 \tag{9.4}$$

for all constant pressure surfaces S. But since ω is proportional to vertical velocity which is proportional to latent heating (see (6.44)),

$$\int_S \dot{Q}_{\mathrm{LH}}\,\mathrm{d}S = 0. \tag{9.5}$$

In other words, to a first approximation, latent heating balances zonally and therefore ENSO heating anomalies are nearly zonally asymmetric.

9.4. AN ATMOSPHERIC ENSO MODEL

Although the ENSO heating anomalies are nearly zonally asymmetric, calculations show that there is a small zonally symmetric heating during El Niño and a small zonally symmetric cooling during La Niña. Apparently the tropical atmosphere is extremely sensitive to such small zonally symmetric heating, because there is a large zonally symmetric warming during El Niño and a large zonally symmetric cooling during La Niña (see Fig. 9.2, left panel). Why should this be so? To address this question it will be helpful to examine the dynamics of a linear atmospheric model similar to that of Gill (1980). The analysis closely follows that of Clarke and Kim (2005).

9.4.1. Logarithmic pressure coordinates

We formulate the model in logarithmic pressure coordinates (x, y, z_*) where here z_* is defined as

$$z_* = -H \ln(p/p_0). \tag{9.6}$$

In (9.6) p_0 is a (constant) pressure surface as close as possible to the surface, p is the atmospheric pressure and the scale height (see also (6.52))

$$H = RT_0/g \tag{9.7}$$

with $T_0 \approx 250\,\text{K}$ being an average virtual temperature (see (6.16)) in the tropical troposphere. To find the relationship of z_* to z, differentiate (9.6) with respect to z and use hydrostatic balance to obtain

$$\frac{\partial z_*}{\partial z} = -\frac{H}{p} p_z = \frac{Hg\rho}{p}. \tag{9.8}$$

It then follows from the equation of state (6.24) and (9.7) that (9.8) can be reduced to

$$\frac{\partial z_*}{\partial z} = \frac{T_0}{T_\nu}. \tag{9.9}$$

Since T_ν is within 25% of T_0 and T_ν averaged over the tropical troposphere is equal to T_0,

$$z_* = \int_0^z T_0/T_\nu \, d\zeta \approx \int_0^z d\zeta = z. \tag{9.10}$$

More accurately, since $T_\nu > T_0$ in approximately the lower half of the troposphere and then $T_\nu < T_0$ in approximately the upper half, the maximum difference between z_* and z occurs at about half the height of the tropical troposphere where it is about 560 m.

9.4.2. Equations of motion in logarithmic pressure coordinates

When we allow a frictional boundary layer near the surface, the horizontal momentum equations can be written in terms of pressure coordinates as (see (6.65), (6.67) and (6.57)):

$$\frac{D\mathbf{u}}{Dt} + f\mathbf{k} \times \mathbf{u} = -\nabla\Phi - g\mathbf{X}_p. \tag{9.11}$$

In (9.11) $\mathbf{X}$ is the turbulent stress near the surface resulting from friction there. Using the definition of z_* in (9.6), (9.11) can be written as

$$\frac{D\mathbf{u}}{Dt} + f\mathbf{k} \times \mathbf{u} = -\nabla\Phi + gH\mathbf{X}_{z_*}/p. \qquad (9.12)$$

To write the continuity equation (6.64) in logarithmic pressure coordinates we must first find the vertical velocity w_* in terms of ω. Since

$$w_* = \frac{Dz_*}{Dt} \qquad (9.13)$$

we have, from (9.6), that

$$w_* = -\frac{H}{p}\frac{Dp}{Dt} = -H\omega/p. \qquad (9.14)$$

Thus from (9.14) the continuity equation (6.64) can be written as

$$\nabla \cdot \mathbf{u} + \frac{\partial}{\partial p}(-pw_*/H) = 0. \qquad (9.15)$$

From (9.6)

$$(w_*)_p = (w_*)_{z_*}(z_*)_p = -(w_*)_{z_*}H/p, \qquad (9.16)$$

so (9.15) reduces to

$$\nabla \cdot \mathbf{u} + (w_*)_{z_*} - w_*/H = 0. \qquad (9.17)$$

The thermodynamic energy equation (6.31) can be written, using (9.14), as

$$\left(\frac{\partial}{\partial t} + \mathbf{u} \cdot \nabla\right)(c_p T(1 + 0.8375q)) + w_*\frac{\partial}{\partial z_*}(c_p T(1 + 0.8375q)) + pw_*/H\rho = \dot{Q}. \qquad (9.18)$$

In the tropics the main balance in (9.18) is between the cooling of a parcel as it rises and the diabatic heating (see (6.44) and the accompanying physical explanation). The specific humidity terms are also small enough to be neglected for our purposes so (9.18) reduces to

$$w_*(T_{z_*} + p/(H\rho c_p)) = \dot{Q}/c_p \qquad (9.19)$$

or, using the equation of state for dry air ((6.15b) with $q = 0$),

$$w_*(T_{z_*} + RT/(Hc_p)) = \dot{Q}/c_p.$$ (9.20)

This may also be written as

$$w_* N_*^2 H/R = \dot{Q}/c_p$$ (9.21)

where

$$N_*^2 = (T_{z_*} + RT/Hc_p)RH^{-1}.$$ (9.22)

The final governing equation, based on hydrostatic balance and the equation of state for dry air, can be found from (9.9) with $T_v = T$. We have

$$\frac{\partial z_*}{\partial z} = T_0/T$$ (9.23)

or, dividing by g, inverting and using (9.7),

$$\frac{\partial \Phi}{\partial z_*} = gT/T_0 = RT/H.$$ (9.24)

9.4.3. Model equations

In order to assess the atmospheric response to ENSO heating, we will analyze the simple case when the ENSO response is a linear perturbation to a state of rest. Since also the ENSO atmospheric response occurs on a time scale short compared to ENSO time scales, the response is quasi-steady and we may take $(\partial/\partial t) = 0$. Denoting perturbation quantities with a prime, the governing equations for the perturbation flow on the equatorial β-plane are, from (9.12), (9.17), (9.24) and (9.21):

$$-\beta y v' = -\Phi'_x + gHX'_{z_*}/p,$$ (9.25)

$$\beta y u' = -\Phi'_y + gHY'_{z_*}/p,$$ (9.26)

$$u'_x + v'_y + \frac{\partial w'_*}{\partial z_*} - w'_*/H = 0,$$ (9.27)

$$\frac{\partial \Phi'}{\partial z_*} = RT'/H$$ (9.28)

and

$$w'_* N^2_* H/R = (\dot{Q})'/c_p. \tag{9.29}$$

We will write the anomalous diabatic heating in the form

$$(\dot{Q})'/c_p = (\dot{Q}_{LH})'/c_p - K_T T' \tag{9.30}$$

where $(\dot{Q}_{LH})'$ is the anomalous latent heating due to condensation and K_T^{-1} is a Newtonian cooling time which has a value of ~ 1 month. Using (9.30) in (9.29) gives

$$K_T T' + w'_* N^2_* H/R = (\dot{Q}_{LH})'/c_p. \tag{9.31}$$

With regard to the near-surface friction, since the winds are mainly zonal $Y'_{z_*} \lesssim X'_{z_*}$. From this and (9.25)

$$gHY'_{z_*}/p \lesssim gHX'_{z_*}/p \lesssim \Phi'_x. \tag{9.32}$$

But zonal scales are much greater than meridional so $\Phi'_x \ll \Phi'_y$ and therefore

$$gHY'_{z_*}/p \ll \Phi'_y. \tag{9.33}$$

Hence (9.26) reduces to

$$\beta y u' = -\Phi'_y, \tag{9.34}$$

a result analogous to (5.7).

The anomalous eastward stress X' is concentrated near the surface and we write

$$X' = X'(0) \exp(-z_*/h) \tag{9.35}$$

where $X'(0)$ is the anomalous eastward turbulent surface stress and h is the frictional boundary layer scale. The surface wind stress depends on a wind speed–dependent drag coefficient c_D, the 10 m height wind speed s and the zonal velocity. A linearized form of $X'(0)$ is usually reasonable (see Wright and Thompson 1983; Clarke et al. 1998) and we simply write

$$X'(0) = \rho(0)c_D s\, u'(0) \tag{9.36}$$

with

$$sc_D \approx 5.2 \times 10^{-3}\,\mathrm{m\,s^{-1}} \tag{9.37}$$

for the typical values $c_D = 1.3 \times 10^{-3}$ and $s = 4\,\text{m}\,\text{s}^{-1}$. A reasonable value for h is about 1 km (Clarke and Lebedev 1997).

Substitution of (9.35) and (9.36) into (9.25) gives

$$\mu u'(0)h^{-1} \exp(-z_*/h) - \beta y v' + \Phi'_x = 0 \tag{9.38}$$

where

$$\mu = gHsc_D\rho(0)/p. \tag{9.39}$$

Since $p \approx$ constant in the lowest kilometer of the troposphere we have

$$\mu = 4.9 \times 10^{-3}\,\text{m}\,\text{s}^{-1} \tag{9.40}$$

using the reasonable values $p = 940\,\text{hPa}$, $\rho(0) = 1.2\,\text{kg}\,\text{m}^{-3}$, $g = 9.8\,\text{m}\,\text{s}^{-2}$, $H = 7.6\,\text{km}$ and (9.37).

Two governing equations in terms of the dependent variables u', v' and Φ' are (9.38) and (9.34). A third equation in these variables can be found by first eliminating T' from (9.28) and (9.31) to obtain

$$w'_* = \frac{-K_T \Phi'_{z_*}}{N_*^2} + \frac{R}{\left(HN_*^2 c_p\right)}(\dot{Q}_{LH})' \tag{9.41}$$

and then substituting for w'_* in (9.27). Taking $N_*^2 = $ constant we obtain

$$u'_x + v'_y - K_T\left(\frac{\Phi'_{z_* z_*}}{N_*^2} - \frac{\Phi'_{z_*}}{HN_*^2}\right) = \left(\frac{\partial}{\partial z_*} - \frac{1}{H}\right)\left(\frac{-(\dot{Q}_{LH})' R}{c_p HN_*^2}\right). \tag{9.42}$$

We will analyze the governing equations (9.38), (9.34) and (9.42) by using vertical modes.

9.4.4. Separation of the governing equations into vertical modes

Expand the dependent variables u', v' and Φ' as follows:

$$u' = \sum_{n=0}^{\infty} u_n(x, y, t) G_n(z_*) \exp(z_*/2H), \tag{9.43}$$

$$v' = \sum_{n=0}^{\infty} v_n(x, y, t) G_n(z_*) \exp(z_*/2H), \tag{9.44}$$

and

$$\Phi' = \sum_{n=0}^{\infty} \phi_n(x, y, t) G_n(z_*) \exp(z_*/2H). \tag{9.45}$$

In (9.43)–(9.45) the vertical mode eigenfunctions are

$$G_0(z_*) = \exp(-z_*/2H) \tag{9.46a}$$

$$G_n(z_*) = \cos\left(\frac{n\pi z_*}{D}\right) - \frac{D}{2n\pi H}\sin\left(\frac{n\pi z_*}{D}\right) \quad n = 1, 2, \ldots \tag{9.46b}$$

where D is the height of the tropopause. These eigenfunctions satisfy the Sturm–Liouville problem

$$\frac{d^2 G_n}{dz_*^2} + \lambda_n G_n = 0 \tag{9.47}$$

$$\frac{dG_n}{dz_*} + \frac{G_n}{2H} = 0 \quad \text{on } z = 0, D \tag{9.48}$$

with eigenvalues

$$\lambda_0 = -1/(4H^2) \tag{9.49}$$

$$\lambda_n = (n\pi/D)^2 \quad n = 1, 2, \ldots. \tag{9.50}$$

The eigenfunctions are orthogonal and form a complete set for $0 \le z_* \le D$ under the boundary conditions (9.48) with orthogonality relationship

$$\int_0^D G_i G_j = \delta_{ij} d_j \tag{9.51}$$

where $\delta_{ij} = 1$ for $i = j$, zero for $i \ne j$ and

$$d_0 = H(1 - \exp(-D/H)), \tag{9.52}$$

$$d_j = \frac{D}{2}\left[1 + \left(\frac{D}{2j\pi H}\right)^2\right] \quad j \ge 1. \tag{9.53}$$

Note from (9.45) and (9.28) that the eigenfunction boundary conditions (9.48) are equivalent to

$$\Phi'_{z_*} \quad \text{and} \quad T' = 0 \quad \text{on } z_* = 0, D. \tag{9.54}$$

The condition at $z_* = D$ is satisfied in the tropics provided that $D \approx 18\,\mathrm{km}$ (Clarke and Kim 2005). Regarding the condition at $z_* = 0$, we will focus on that part of the temperature anomaly driven by anomalous deep atmospheric convection. For such heating $(\dot{Q}_{LH})' = 0$ at the surface $z_* = 0$. Since also $w'_* = 0$ at the surface, it follows from (9.31) that the temperature anomaly vanishes at the surface, consistent with (9.54).

Substitution of (9.43)–(9.45) into (9.38) and then multiplying through by $G_i(z_*) \exp(-z_*/2H)$ gives, upon use of the orthogonality relation (9.51), the vertical mode i zonal momentum equation

$$k_i u_i + k_i \sum_{\substack{j=0 \\ j \neq i}}^{\infty} u_j - \beta y v_i + \frac{\partial \phi_i}{\partial x} = 0 \tag{9.55}$$

where the k_i are all given by

$$k_i = \frac{\mu}{h d_i} \int_0^D G_i \exp(-z_*/h)\exp(-z_*/2H)\mathrm{d}z_*. \tag{9.56}$$

Similarly, substituting (9.43)–(9.45) into (9.34) and (9.42) gives two other sets of equations, analogous to (9.55), for the horizontally varying functions u_i, v_i and ϕ_i. In deriving the second of these sets of equations from (9.42) we use

$$\frac{\mathrm{d}^2}{\mathrm{d}z_*^2}[G_j \exp(z_*/2H)] - \frac{1}{H}\frac{\mathrm{d}}{\mathrm{d}z_*}[G_j \exp(-z_*/2H)]$$

$$= \begin{cases} \dfrac{-N_*^2}{c_j^2} G_j \exp(-z_*/2H) & j = 1, 2, \dots & \text{(9.57a)} \\ 0 & j = 0 & \text{(9.57b)} \end{cases}$$

where

$$c_j^2 = N_*^2/(j^2\pi^2/D^2 + 1/(4H^2)). \tag{9.58}$$

The two sets of equations are

$$\beta y u_i = -\frac{\partial \phi_i}{\partial y} \tag{9.59}$$

and

$$(1 - \delta_{i0})\frac{K_T \phi_i}{c_i^2} + \frac{\partial u_i}{\partial x} + \frac{\partial v_i}{\partial y} = -P_i \tag{9.60}$$

where

$$P_i = \frac{1}{d_i} \int_0^D G_i \exp(-z_*/2H) \left(\frac{d}{dz_*} - \frac{1}{H} \right) (\dot{Q}_{LH})' R/(c_p N_*^2 H) dz_* \quad (9.61)$$

and δ_{i0} is zero except when $i = 0$ when it is unity.

Analytical expressions for k_i, the inverse of the frictional decay time, can be obtained for the lower order vertical modes. These modes and $2H$ have a vertical scale much larger than h and so, in (9.56), $G_i(z_*) \exp(-z_*/2H)$ may be approximated by its value at $z_* = 0$. Thus, from (9.56), (9.46) and (9.53) we have

$$k_i \approx \frac{\mu}{hd_i} G_i(0) \int_0^D \exp(-z_*/h) dz_* \approx \mu/d_i \approx 2\mu/D \quad i \neq 0 \qquad (9.62)$$

and

$$k_0 \approx \mu/d_i = \mu/(H(1 - \exp(-D/H))). \qquad (9.63)$$

Frictional decay time scales can be estimated using (9.62), (9.63), (9.40), $H = 7.6$ km and $D = 18$ km:

$$(k_i)^{-1} \approx 21 \text{ days for } i \text{ small and non-zero}$$

$$(k_0)^{-1} \approx 16 \text{ days.} \qquad (9.64)$$

Clarke and Kim (2005) suggest that the other dissipative time scale, the Newtonian cooling time $K_T^{-1} \approx 1$ month.

In summary, the solution of the atmospheric model is given by (9.43)–(9.45). The functions $u_i(x, y, t)$, $v_i(x, y, t)$ and $\phi_i(x, y, t)$ satisfy (9.55), (9.59) and (9.60) and the dissipation time scales $(k_i)^{-1}$ and $(K_T)^{-1}$ are known.

9.4.5. Dominance of the first vertical mode

As pointed out by Gill (1980) for a simpler model, vertical mode 1 dominates the response in the tropics (see Fig. 9.6) . We can see why this is so from our model by first noting that $\dot{Q}'_{LH}$ for deep convective ENSO heating (see Fig. 9.7) tends to be of one sign in the vertical with a maximum in the mid to upper troposphere, i.e., it looks like $\exp(z_*/2H) \sin(\pi z_*/D)$. But by (9.42), the anomalous heating is proportional to the forcing term

$$-\left(\frac{\partial}{\partial z_*} - \frac{1}{H} \right) \dot{Q}'_{LH}$$

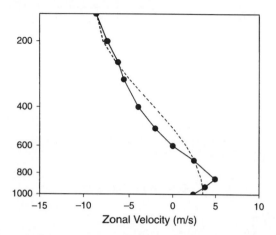

Fig. 9.6 El Niño composite since 1970 of the vertical structure of the equatorial zonal wind anomaly in December (solid line with black dots) averaged over the central Pacific region (180°E–230°E) from NCEP–NCAR Reanalysis Data. The El Niño years are 1972, 1976, 1982, 1987, 1991 and 1997. The ordinate represents vertical level in hPa using a logarithmic scale and the abscissa represents zonal velocity in m s^{-1}. The dashed curve is the first vertical mode structure function $\exp\left(\frac{z}{2H}\right)\left[\cos\left(\frac{\pi z}{D}\right) - \frac{D}{2\pi H}\sin\left(\frac{\pi z}{D}\right)\right]$. (From Clarke and Kim 2005.)

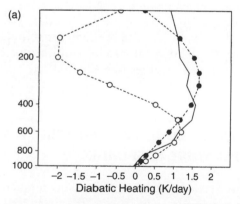

Fig. 9.7 El Niño composite (solid line) of $\dot{Q}'_{LH}/c_p$ (residually diagnosed diabatic heating from the NCEP/NCAR Reanalysis Data) for the region (170°W–130°W, 10°S–5°N) in units of °K day^{-1}. Also shown are the relevant vertical velocity structure functions $\exp(z/2H)\sin(j\pi z/D)$ for the $j=1$ (solid circles) and $j=2$ (open circles) vertical modes. The latter are in dimensionless units but the abscissa is numerically the same for all plots. The El Niño composite is based on the December–February value of 1982–1983, 1987–1988, 1991–1992 and 1997–1998. (b) As for (a) but for La Niña using the December–February values of 1984–1985, 1988–1989, 1995–1996, 1998–1999 and 1999–2000. (From Clarke and Kim 2005.)

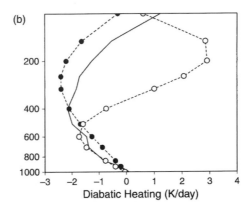

(b)

Diabatic Heating (K/day)

Fig. 9.7   Continued.

which has an approximate vertical structure

$$\left(\frac{d}{dz_*} - \frac{1}{H}\right)\{\exp(z_*/2H)\sin(\pi z_*/D)\} = \frac{\pi}{D}\exp(z_*/2H)G_1(z_*) \quad (9.65)$$

with $G_1(z_*)$ defined as in (9.46b). Thus by (9.61), (9.65) and the orthogonality relation for the eigenfunctions, the forcing term P_i should be much larger in magnitude for $i = 1$ than the other modes. We might therefore expect vertical mode one to dominate the response. Calculations with realistic ENSO forcing $\dot{Q}'_{LH}$ using the model confirm this even though the vertical modes are coupled together (see (9.55)).

9.5. THE LARGE ZONALLY SYMMETRIC AIR TEMPERATURE RESPONSE

As discussed in Section 9.2, much of the air temperature response to ENSO is zonally symmetric even though the deep convective heating is nearly zonally asymmetric. We can understand this physically (Wu et al. 2001; Clarke and Kim 2005) by considering the model of Section 9.4 under the approximation that the response is dominated by the first vertical mode (Section 9.4.5).

If all the response is in the first vertical mode then the governing equations for that mode are (9.59), (9.60) and (9.55) with the second term on the left-hand side missing. These equations are of the same form as the frictional surface boundary layer model of Chapter 6 (see the zonal momentum component of (6.73) together with the dimensional version of (6.87) and (6.80)). In our present case the momentum damping time scale $k_1^{-1} \approx 21$ days is much longer than the momentum damping time scale $K^{-1} = 2 \times 10^5 \, \mathrm{s} \approx 2$ days in (6.73).

As shown in Chapter 6, the governing equations can be written in the non-dimensional form (6.81), (6.87) and (6.83). The solution of these equations is (6.89)–(6.94). Physically, these equations describe a sum of a damped, forced long equatorial Kelvin wave and damped, forced, long equatorial Rossby waves. It follows from (6.92) that, in the absence of forcing, the amplitude q_0 of the equatorial Kelvin wave satisfies the non-dimensional equation

$$K' q_0 + \frac{\partial q_0}{\partial x'} = 0. \tag{9.66}$$

The solution of (9.66) decays eastward exponentially with a non-dimensional scale $(K')^{-1}$ or, in the notation of Chapter 6, a dimensional scale $(K')^{-1} (c/\beta)^{1/2} (K/K_T)^{1/4}$. Using (6.84) and replacing K by k_1 (as is appropriate for the present chapter), we have a dimensional zonal decay scale of $c/(K_T k_1)^{1/2}$. Since the first vertical mode dominates the response, c is given by (9.58) with $j = 1$. Using the reasonable values $c = 68 \, \mathrm{m \, s^{-1}}$, $(K_T)^{-1} \approx 30$ days and $(k_1)^{-1} \approx 21$ days, the decay scale is

$$c/(k_1 K_T)^{1/2} \approx 147\,000 \, \mathrm{km}, \tag{9.67}$$

more than three times the circumference of the earth. This implies that any locally forced equatorial Kelvin wave will decay only slightly around the earth, i.e., the Kelvin wave contribution to the response will have a large zonally symmetric component. The first meridional mode Rossby wave also contributes substantially to the ENSO response and, from (6.93), when unforced it decays westward with a scale $c/[(k_1 K_T)^{1/2} (2m+1)]$ with $m = 1$. From (9.67) this scale is $49\,000$ km, again longer than the circumference of the earth. Consequently its contribution to the response will also have a substantial zonally symmetric component. Thus the zonally symmetric response is large because the damping is light for the lower order modes which contribute substantially to the response. The equatorial Kelvin wave with speed c and first meridional mode Rossby wave with speed $c/3$ propagate at least once around the earth during the dissipation time scale $(k_1 K_T)^{-1/2} = 25$ days.

9.6. MID-LATITUDE ZONALLY SYMMETRIC COOLING DURING WARM ENSO EVENTS

Figure 9.4 shows that while the tropical 200 hPa heights are anomalously high and therefore the tropical troposphere is warmer than normal, the mid-latitude troposphere tends to have lower than normal 200 hPa heights and be cooler than normal. Much of this variability is zonally symmetric and in this section we will examine this tropical–mid-latitude zonally symmetric 'teleconnection.' The cooler than normal signal is due mainly to the influence of eddies on the large-scale low-frequency flow so we will first examine the large-scale eddy-driven equations next and then discuss the cooling following the theory of Seager et al. (2003). Note that these warm ENSO event effects reverse during a cold ENSO event; then the tropical troposphere is cooler than normal and the mid-latitude troposphere tends to be anomalously warm.

Our focus is in mid-latitudes where, at lowest order, the flow is geostrophic. For this small Rossby number flow

$$\nabla \cdot \mathbf{u} \approx 0 \quad \text{and} \quad \frac{D}{Dt} \approx \frac{\partial}{\partial t} + \mathbf{u} \cdot \nabla \qquad (9.68)$$

where $\mathbf{u}$ is the horizontal velocity. Friction does not play an important role in the dynamics to be discussed so we may write the governing zonal momentum equation in (9.11) as

$$\frac{\partial u}{\partial t} + \mathbf{u} \cdot \nabla u - fv = -\Phi_x. \qquad (9.69)$$

Using (9.68) this equation is equivalent to

$$\frac{\partial u}{\partial t} + \nabla \cdot (\mathbf{u}u) - fv = -\Phi_x. \qquad (9.70)$$

Now write

$$\mathbf{u} = \bar{\mathbf{u}} + \mathbf{u}'', \quad \Phi = \bar{\Phi} + \Phi'' \qquad (9.71)$$

where the overbar denotes a monthly average and the double prime the departure from that monthly average. Since $\overline{\mathbf{u}''}$ and $\overline{\Phi''}$ vanish, substitution of (9.71) into (9.70) and then taking the monthly average gives

$$\frac{\partial \bar{u}}{\partial t} + \nabla \cdot (\bar{\mathbf{u}}\bar{u}) + \nabla \cdot (\overline{\mathbf{u}''u''}) - f\bar{v} = -\bar{\Phi}_x. \qquad (9.72)$$

Subdivide the monthly averaged variables further by writing

$$\bar{u} = [\bar{u}] + \bar{u}_{**}, \quad \bar{\Phi} = [\bar{\Phi}] + \bar{\Phi}_{**} \tag{9.73}$$

where the square bracket denotes a zonal average and double asterisk the departure from that zonal average. Since $[\bar{u}_{**}]$ and $[\bar{\Phi}_{**}]$ vanish, substitution of (9.73) into (9.72) and then taking the zonal average gives

$$\frac{\partial [\bar{u}]}{\partial t} + \frac{\partial}{\partial y}[\bar{v}][\bar{u}] + \frac{\partial}{\partial y}[\bar{v}_{**}\bar{u}_{**}] + \frac{\partial}{\partial y}[\overline{u''v''}] = f[\bar{v}]. \tag{9.74}$$

Seager et al. (2003) found that, for the anomaly version of (9.74), the first four terms on the left-hand side were dominated by the divergence of the anomalous transient eddy flux, i.e.,

$$\frac{\partial}{\partial y}[\overline{u''v''}]' = f[\bar{v}]' \tag{9.75}$$

where the single prime refers to the anomaly. Physically, the divergence of the zonally averaged transient eddy flux anomaly is balanced by the anomalous Coriolis force. In this way the transient eddy flux anomaly drives a meridional zonally averaged low-frequency flow anomaly.

Two other equations are necessary to understand the anomalous mid-latitude cooling seen during a warm ENSO event. One is found by monthly and zonally averaging the continuity equation (9.17) and then subtracting out the mean and seasonal cycle to obtain

$$\exp(-z_*/H)\frac{\partial}{\partial y}[\bar{v}]' + \frac{\partial}{\partial z_*}([\bar{w}_*]'\exp(-z_*/H)) = 0. \tag{9.76}$$

The second equation results from an analysis of the non-linear thermodynamic energy equation in the same way we analyzed the zonal momentum equation. Eddy fluxes result, but for the interannual variability of interest, the key terms are vertical advection and Newtonian cooling. Thus the appropriate interannual equation is the monthly and zonally averaged version of (9.31) with $(\dot{Q}_{LH})'$ ignored:

$$K_T[T]' = -[\bar{w}_*]'N_*^2 H/R. \tag{9.77}$$

An analysis of the anomalous mid-latitude cooling can now be given. As discussed earlier in this chapter, during a warm ENSO event the tropical troposphere warms anomalously and the geopotential height is higher than normal. By geostrophic balance, westerly wind anomalies at the edge of the tropics strengthen the subtropical atmospheric jets on their equatorward flanks at about 20° latitude in both hemispheres. This alters the background state on which transient eddies propagate and hence also the transient eddy zonal momentum flux and its divergence $(\partial/\partial y)[\overline{u''v''}]'$. The eddy flux convergence $-(\partial/\partial y)[\overline{u''v''}]'$ in the upper troposphere applies a westerly (eastward) force between about latitudes 30° and 45° in both hemispheres and an easterly (westward) force between about 45° and 60° in both hemispheres. These forces generate an anomalous upper tropospheric meridional flow $[\bar{v}]'$ so that the eddy flux convergence is balanced by the Coriolis force $-f[\bar{v}]'$ (see (9.75) multiplied by -1). As a result, in both hemispheres upper tropospheric $[\bar{v}]'$ is poleward at latitudes higher than 45° and equatorward at latitudes lower than 45°. This upper tropospheric divergence at 45° latitude and $[\bar{w}_*]' = 0$ at the tropopause implies that $[\bar{w}_*]' > 0$ in the upper troposphere in mid-latitudes. The rising air cools anomalously under the lowest order balance (9.77) and there is thus anomalous mid-latitude upper tropospheric cooling.

➤ 9.7. ZONALLY ASYMMETRIC MID-LATITUDE TELECONNECTIONS

9.7.1. Introduction

Figures 9.4 and 9.5 show that, while the atmospheric response to ENSO heating has a strong zonally symmetric component, there are local anomalous features. For example, in the tropics (Fig. 9.4) the 200 mb height anomalies form a dumbbell shape at about 140°W that Fig. 9.4(c) and (d) shows is a clear and strong signal of opposite sign during El Niño and La Nina. There is also a clear and strong extratropical signal in the Gulf of Alaska which, near its center (see the dot in Fig. 9.5), is of opposite sign during El Niño and La Niña.

Why should there be remote regions where the ENSO signal is enhanced? Much research has been done on ENSO teleconnections, particularly from the equatorial Pacific to the extratropical Pacific and North America. Here we will concentrate on some of the basic ideas in the theory of why certain extratropical regions see a strong ENSO signal while others do not. Early theoretical work on this problem by Opsteegh and van den Dool (1980), Webster (1981) and Hoskins and Karoly (1981) was first applied to observations by Horel and Wallace (1981). Our treatment will focus on the most basic aspects of the theory and will draw on elements of Trenberth et al. (1998) and Holton (2004).

9.7.2. Stationary Rossby wave theory

From (9.12) the horizontal momentum equations in logarithmic pressure coordinates are

$$\frac{\partial \mathbf{u}}{\partial t} + \mathbf{u} \cdot \nabla \mathbf{u} + w_* \frac{\partial \mathbf{u}}{\partial z_*} + f\mathbf{k} \times \mathbf{u} = -\nabla \Phi + gH\mathbf{X}_{z_*}/p \qquad (9.78)$$

where $\mathbf{u}$ is the horizontal velocity and ∇ is the horizontal gradient operator. Surface friction does not play an essential role in the dynamics to be discussed so we omit the last term on the right-hand side of (9.78). We also use the vector identity

$$\mathbf{u} \cdot \nabla \mathbf{u} = \nabla \left(\frac{1}{2} |\mathbf{u}|^2 \right) - \mathbf{u} \times (\nabla \times \mathbf{u}) \qquad (9.79)$$

where

$$\nabla \times \mathbf{u} = \zeta \mathbf{k} \qquad (9.80)$$

and the vertical component of relative vorticity

$$\zeta = \frac{\partial v}{\partial x} - \frac{\partial u}{\partial y}. \qquad (9.81)$$

Substitution of (9.79) and (9.80) into the frictionless version of (9.78) gives

$$\frac{\partial \mathbf{u}}{\partial t} + \nabla \left(\frac{1}{2} |\mathbf{u}|^2 \right) + w_* \frac{\partial \mathbf{u}}{\partial z_*} + (\zeta + f)\mathbf{k} \times \mathbf{u} = -\nabla \Phi. \qquad (9.82)$$

Taking $\nabla \times$ of (9.82) gives the vorticity equation

$$\frac{\partial(\zeta \mathbf{k})}{\partial t} + (\nabla w_*) \times \frac{\partial \mathbf{u}}{\partial z_*} + w_* \frac{\partial(\zeta \mathbf{k})}{\partial z_*} + \nabla(\zeta + f) \times (\mathbf{k} \times \mathbf{u}) + (\zeta + f)\nabla \times (\mathbf{k} \times \mathbf{u}) = 0. \qquad (9.83)$$

Since $\mathbf{k}$ is vertical and constant and ∇ is horizontal, we have, from the vector identities

$$\nabla(\zeta + f) \times (\mathbf{k} \times \mathbf{u}) = \mathbf{k}(\mathbf{u} \cdot \nabla(\zeta + f)) - \mathbf{u}(\mathbf{k} \cdot \nabla(\zeta + f)) \qquad (9.84)$$

and

$$\nabla \times (\mathbf{k} \times \mathbf{u}) = \mathbf{k}(\nabla \cdot \mathbf{u}) + \mathbf{u} \cdot \nabla(\mathbf{k}) - \mathbf{u}\nabla \cdot \mathbf{k} - \mathbf{k} \cdot \nabla u \qquad (9.85)$$

that (9.83) can be written as

$$\frac{\partial(\zeta\mathbf{k})}{\partial t} + (\boldsymbol{\nabla}w_*) \times \frac{\partial\mathbf{u}}{\partial z_*} + w_*\frac{\partial(\zeta\mathbf{k})}{\partial z_*} + \mathbf{k}(\mathbf{u}\cdot\boldsymbol{\nabla}(\zeta+f)) + (\zeta+f)\mathbf{k}(\boldsymbol{\nabla}\cdot\mathbf{u}) = 0. \tag{9.86}$$

Taking the dot product of (9.86) with $\mathbf{k}$ and using (9.17) then gives

$$\frac{\partial\zeta}{\partial t} + \mathbf{u}\cdot\boldsymbol{\nabla}(\zeta+f) + \mathbf{k}\cdot\boldsymbol{\nabla}w_* \times \frac{\partial\mathbf{u}}{\partial z_*} + w_*\frac{\partial\zeta}{\partial z_*} = (\zeta+f)\left[\frac{\partial w_*}{\partial z} - \frac{w_*}{H}\right]. \tag{9.87}$$

If we analyze (9.87) near the tropopause where $w_* = 0$ then the vorticity equation reduces to

$$\frac{\partial\zeta}{\partial t} + \mathbf{u}\cdot\boldsymbol{\nabla}(\zeta+f) = (\zeta+f)\frac{\partial w_*}{\partial z_*} \tag{9.88}$$

or, using the relationship (9.21) between vertical velocity and the rate of heating per unit mass,

$$\frac{\partial\zeta}{\partial t} + \mathbf{u}\cdot\boldsymbol{\nabla}(\zeta+f) = (\zeta+f)(\dot{Q}R/(c_pN_*^2H))_{z_*}. \tag{9.89}$$

The time mean zonal flow $\bar{u}$ has a profound influence on ENSO teleconnections. To understand some basic properties of this influence we linearize the vorticity equation (9.89) about a zonal mean flow $\bar{u}$ which is independent of x to obtain

$$\frac{\partial\zeta'}{\partial t} + \bar{u}\frac{\partial}{\partial x}\zeta' + v'\frac{\partial}{\partial y}(-\bar{u}_y+f) = \left[(\zeta+f)(\dot{Q}R/(c_pN_*^2H))_{z_*}\right]'. \tag{9.90}$$

At this point in basic teleconnection theory it is customary to introduce a stream function ψ with

$$v' = \psi'_x, \quad u' = -\psi'_y. \tag{9.91}$$

However, (9.91) implies that the flow is at least approximately horizontally non-divergent, but near the equator where there is strong anomalous heating, the flow is horizontally divergent. Consequently (9.91) can only be used away from the equator where the flow is quasi-geostrophic and so approximately horizontally non-divergent. Substitution of (9.91) into (9.90) gives

$$\left(\frac{\partial}{\partial t} + \bar{u}\frac{\partial}{\partial x}\right)\nabla^2\psi' + (\beta - \bar{u}_{yy})\psi'_x = \left[(\zeta+f)(\dot{Q}R/(c_pN_*^2H))_{z_*}\right]'. \tag{9.92}$$

Away from the region of strong anomalous heating, the right-hand side of (9.92) is negligible and can be omitted. If $\bar{u}$ and $\bar{u}_{yy}$ vary 'slowly' in y compared to variations in ψ' then $\bar{u}$ and $\bar{u}_{yy}$ can be regarded as locally constant. In that case the solution to the homogeneous version of (9.92) is of the form

$$\psi' = A\cos(kx + ly - \sigma t) \tag{9.93}$$

provided that the frequency σ satisfies the Rossby wave dispersion relation

$$\sigma = \bar{u}k - (\beta - \bar{u}_{yy})k/(k^2 + l^2). \tag{9.94}$$

From (9.93) stationary wave patterns of the form $A\cos(kx + ly)$ are possible when $\sigma = 0$. The amplitude and sign of these patterns will vary spatially according to $\cos(kx + ly)$; there will be strong teleconnections of opposite sign when $kx + ly$ are even and odd multiples of π and no indication of teleconnections for those points where $kx + ly$ is an odd multiple of $\pi/2$.

To analyze the Rossby wave pattern in more detail we must calculate the group velocity, for this vector will tell us where the energy goes. The group velocity is

$$\mathbf{c_g} = c_{gx}\mathbf{i} + c_{gy}\mathbf{j} \tag{9.95}$$

with the zonal and meridional group velocity components c_{gx} and c_{gy} being found from (9.94) as

$$c_{gx} = \frac{\partial\sigma}{\partial k} = \bar{u} + (\beta - \bar{u}_{yy})(k^2 - l^2)/(k^2 + l^2)^2 \tag{9.96a}$$

and

$$c_{gy} = \frac{\partial\sigma}{\partial l} = 2(\beta - \bar{u}_{yy})kl/(k^2 + l^2)^2. \tag{9.96b}$$

For stationary wave patterns $\sigma = 0$ and so, from (9.94) with $\sigma = 0$ we have

$$\bar{u} = (\beta - \bar{u}_{yy})/(k^2 + l^2). \tag{9.97}$$

Thus (9.96a) and (9.96b) can be written as

$$c_{gx} = (2\bar{u}k^2)/(k^2 + l^2) \tag{9.98a}$$

and

$$c_{gy} = (2\bar{u}kl)/(k^2 + l^2). \tag{9.98b}$$

Since $\beta - \bar{u}_{yy}$ is usually positive, (9.97) implies that $\bar{u}$ must also be positive. Physically, a steady solution is only possible when the westward-propagating Rossby wave is in an eastward flow at the same speed as the westward propagation. Mathematically, (9.97) holds when $\bar{u}$ matches the westward Rossby wave propagation speed

$$\sigma/k = -(\beta - \bar{u}_{yy})/(k^2 + l^2). \tag{9.99}$$

Since $\bar{u}$ is positive, from (9.98a) the zonal group velocity is eastward. The meridional component can be positive or negative and the tangent of the angle that the group velocity makes with the zonal direction is, by (9.98a),

$$c_{gy}/c_{gx} = l/k. \tag{9.100}$$

Thus the standing Rossby wave teleconnections will be eastward of the region of forcing and will extend northward and southward of it at an angle $\tan^{-1}(l/k)$ to the horizontal.

A schematic view of how extratropical teleconnections might occur, based on the above theory, is given in Fig. 9.8. The anomalous equatorial divergence in the upper troposphere due to anomalous deep convective heating results in an outflow which generates the anomalous anti-cyclone pair on either side of the equator. The quasi-stationary Rossby wave response to this pair is a series of lows and highs eastward and poleward along the arcing double line shown in Fig. 9.8.

Although the foregoing theory of this section is helpful in understanding extratropical teleconnections, the related often-quoted mechanism summarized by the schematic of Fig. 9.8 is inaccurate in at least two ways. First, the anomalous deep-convective equatorial heating is typically centered at about 170°W–180°W (see Figs 2.2(c) and 2.9(c)) but the anti-cyclone pair is centered about 35° to the east at 140°W (see Fig. 9.4). Second, the biggest and most consistent extratropical response does not arc eastward of the anomalous anticyclones as in the schematic but rather is slightly to the west of them (see Fig. 9.4). One key factor not taken into account by the schematic Fig. 9.8 and the stationary Rossby wave theory discussed in this section is that the observed mean flow varies strongly with longitude. This longitudinal variation has a profound effect on the location and strength of

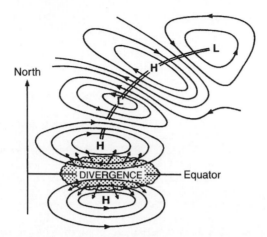

Fig. 9.8 Extratropical teleconnections schematic based on stationary Rossby wave theory. Anomalous equatorial tropical convection (scalloped region) generates anomalous equatorial tropical divergence which results in an outflow that converges and causes the anomalous anticyclones (marked by H) immediately to the north and south of the anomalous convection. A wave train of alternating lows and highs spreads eastward and poleward along in the direction of the group velocity (double line) by the stationary Rossby wave mechanism discussed in the text. (Adapted from Trenberth et al. 1998.)

the teleconnections (see, e.g., Branstator 1983, 1985; Karoly 1983; Webster and Chang 1988; Hoskins and Ambrizzi 1993). In addition, among other things teleconnections are influenced by transient vorticity fluxes (Kok and Opsteegh 1985 and Held et al. 1989) and by anomalous ENSO cooling as well as heating, especially in the western equatorial Pacific (DeWeaver and Nigam 2004).

REFERENCES

Branstator, G., 1983: Horizontal energy propagation in a barotropic atmosphere with meridional and zonal structure. *J. Atmos. Sci.*, **40**, 1689–1708.

Branstator, G., 1985: Analysis of general circulation model sea-surface temperature anomaly simulations using a linear model. Part I: Forced solutions. *J. Atmos. Sci.*, **42**, 2225–2241.

Clarke, A. J., and K. Y. Kim, 2005: On weak zonally symmetric ENSO atmospheric heating and the strong zonally symmetric ENSO air temperature response. *J. Atmos. Sci.*, **62**(6), 2012–2022.

Clarke, A. J., and A. Lebedev, 1997: Interannual and decadal changes in equatorial windstress in the Atlantic, Indian and Pacific Oceans and the eastern ocean coastal response. *J. Clim.*, **10**, 1722–1729.

Clarke, A. J., X. Liu, and S. Van Gorder, 1998: Dynamics of the biennial oscillation in the equatorial Indian and far western Pacific Oceans. *J. Clim.*, **11**(5), 987–1001.

DeWeaver, E., and S. Nigam, 2002: Linearity in ENSO's atmospheric response. *J. Clim.*, **15**, 2446–2461.

DeWeaver, E., and S. Nigam, 2004: On the forcing of ENSO teleconnections by anomalous heating and cooling. *J. Clim.*, **17**, 3225–3235.

Gill, A. E., 1980: Some simple solutions for heat-induced tropical circulation. *Q. J. R. Meteorol. Soc.*, **106**, 447–462.

Halpert, M. S., and C. F. Ropelewski, 1992: Surface temperature patterns associated with the Southern Oscillation. *J. Clim.*, **5**(6), 577–593.

Held, I. M., S. W. Lyons, and S. Nigam, 1989: Transients and the extratropical response to El Niño. *J. Atmos. Sci.*, **46**, 163–174.

Holton, J. R., 2004: *An Introduction to Dynamic Meteorology*, 4th Edition. Elsevier Academic Press, New York, 535 pp.

Horel, J. D., and J. M. Wallace, 1981: Planetary-scale atmospheric phenomena associated with the Southern Oscillation. *Mon. Weather Rev.*, **109**, 813–829.

Hoskins, B. J., and T. Ambrizzi, 1993: Rossby wave propagation on a realistic longitudinally varying flow. *J. Atmos. Sci.*, **50**, 1661–1671.

Hoskins, B. J., and D. J. Karoly, 1981: The steady linear response of a spherical atmosphere to thermal and orographic forcing. *J. Atmos. Sci.*, **38**, 1179–1196.

Karoly, D. J., 1983: Rossby wave propagation in a barotropic atmosphere. *Dyn. Atmos. Oceans*, **7**, 111–125.

Kok, C. J., and J. O. Opsteegh, 1985: On the possible causes of anomalies in seasonal mean circulation pattern during the 1982–83 El Niño event. *J. Atmos. Sci.*, **42**, 677–694.

Opsteegh, J. D., and H. M. Van Den Dool, 1980: Seasonal differences in the stationary response of a linearized primitive equation model: Prospects for long-range forecasting? *J. Atmos. Sci.*, **37**, 2169–2185.

Seager, R., N. Harnik, Y. Kushnir, W. Robinson, and J. Miller, 2003: Mechanisms of hemispherically symmetric climate variability. *J. Clim.*, **16**, 2960–2978.

Trenberth, K. E., G. W. Branstator, D. Karoly, A. Kumar, N.-G. Lau, and C. Ropelewski, 1998: Progress during TOGA in understanding and modeling global teleconnections associated with tropical sea surface temperatures. *J. Geophys. Res.*, **103**, 14291–14324.

Webster, P. J., 1981: Mechanisms determining the atmospheric response to sea surface temperature anomalies. *J. Atmos. Sci.*, **38**, 554–571.

Webster, P. J., and H.-R. Chang, 1988: Equatorial energy accumulation and emanation regions: Impacts of a zonally varying basic state. *J. Atmos. Sci.*, **45**, 803–829.

Wright, D. G., and K. R. Thompson, 1983: Time-averaged forms of the nonlinear stress law. *J. Phys. Oceanogr.*, **13**, 341–345.

Wu, Z., E. S. Sarachik, and D. S. Battisti 2001: Thermally driven tropical circulations under Rayleigh friction and Newtonian cooling: Analytic solutions. *J. Atmos. Sci.*, **58**, 724–741.

Yulaeva, E., and J. M. Wallace, 1994: The signature of ENSO in global temperature and precipitation fields derived from the microwave sounding unit. *J. Clim.*, **7**, 1719–1736.

CHAPTER TEN

ENSO FORECASTING USING DYNAMICAL MODELS

Contents

10.1. OVERVIEW

These days there are many coupled ocean–atmosphere models that forecast El Niño every month. One example is the Lamont intermediate model of Chen et al. (2004), a modified version of the model of Zebiak and Cane (1987), the first dynamical model to make a successful El Niño forecast and the model discussed in Section 7.4. Other forecast models, e.g., the coupled general circulation models listed in Table 10.1, make fewer approximations but are much more complicated and more expensive computationally. They involve the global ocean and global atmosphere instead of just the Pacific and are based on the basic (primitive) equations of geophysical fluid dynamics rather than simplified equations. A third group of coupled forecast models (see, e.g., Barnett et al. 1993; Pierce 1996; Kang and Kug 2000; Zhang et al. 2003) are 'hybrid' models in that they combine a dynamical ocean model with a statistically based atmospheric model.

Table 10.1 Coupled ocean–atmosphere dynamical models and the periods of time their performance was assessed and the reference documenting the assessment.

Model	Model type	Period hindcast	Reference
Lamont-Doherty Earth Observatory (LDEO5)	Intermediate coupled ocean–atmosphere model	1980–2003	Chen et al. (2004)
National Centers for Environmental Prediction (NCEP) Climate Forecast System (CFS)	Coupled ocean–land–atmosphere model (coupled ocean and atmosphere general circulation models)	1981–2004	Saha et al. (2006)
ECHAM-MOM (International Research Institute for Climate Prediction)	Coupled ocean–atmosphere general circulation model	1982–2002	DeWitt (2005)
European Centre for Medium-Range Weather Forecasts (ECMWF)	Coupled ocean–atmosphere general circulation model	1987–2006	Van Oldenborgh et al. (2005)
Center for Ocean–Land–Atmosphere Studies (COLA)	Anomaly coupled general circulation model	1980–1999	Kirtman (2003)
Seoul National University, Korea	Hybrid coupled model (intermediate ocean model/statistical atmosphere model)	1980–1999	Kang and Kug (2000); Kug et al. (2001)

We will not examine in detail all the coupled forecast models with a dynamical component. Rather, in Sections 10.2–10.4, we will focus on ways to overcome a major difficulty commonly faced by dynamical models, namely, how best to use limited imperfect observations to improve the performance of an imperfect model. Many assimilation techniques are illustrated by the Lamont model and in Sections 10.2.2, 10.2.3 and 10.3 we discuss 'bias correction,' 'nudging' and the Kalman filter using this model. Adjoint data assimilation is considered in Section 10.4 and then, in Section 10.5, we will discuss the forecast performance of the operational coupled dynamical models in Table 10.1.

10.2. Smoothing, Bias Correction and Nudging

10.2.1. Smoothing

Empirical orthogonal functions (EOFs) (see Appendix A) can be used to smooth noisy data by truncating the number of EOFs used to represent those data. Such truncations can remove uncorrelated high frequency and small-scale 'noise.' For example, hybrid coupled forecast models minimize the errors in the initial observed wind forcing used to force the dynamical ocean model by truncating the EOFs used to represent the wind forcing.

10.2.2. Bias correction

Bias correction also smoothes the data but, in addition, uses the known error between past model and observed data to correct the model statistically. Such a correction was used by Barnett et al. (1993) on their hybrid model and also by Chen et al. (2000) on the Lamont model. We will discuss the Lamont bias correction below.

To use bias correction, we must first estimate the error between observed and model variables at past times t at each horizontal location (x, y). In practice in the Lamont model these variables are the anomalous SST, wind stress and sea level. As for the hybrid models discussed in Section 10.2.1 above, the Lamont model data and the observed data are smoothed by truncating the number of EOFs used to represent the data. The observed SST was truncated at six modes while the observed wind stress and sea level each used eight EOF modes. While the observed EOF analyses were done separately for each physical variable, the model SST, wind stress and sea level were analyzed together. Their data were equally weighted by normalizing each physical variable by its standard deviation over the whole spatial domain and then a multivariate EOF (MEOF) was calculated. Eight MEOF modes were retained. Upon conversion back to physical variables by multiplying by the appropriate standard deviation, the SST, for example, can be written as

$$\mathrm{SST}_{\mathrm{mod}}(x, y, t) = \sum_{n=1}^{8} b_n(t) s_n(x, y) \tag{10.1}$$

and an error field $E(x, y, t)$ defined by

$$E(x, y, t) = \mathrm{SST}_{\mathrm{mod}}(x, y, t) - \mathrm{SST}_{\mathrm{obs}}(x, y, t). \tag{10.2}$$

Note that while SST_{mod} is represented by eight EOF modes, SST_{obs} is represented by only six. The SST error field $E(x, y, t)$ is then represented by another EOF expansion, truncated to six terms, the same number of terms as the observations:

$$E(x, y, t) \approx \sum_{n=1}^{6} a_n(t) e_n(x, y).$$ (10.3)

While the error EOFs $e_n(x, y)$ in (10.3) are available to correct SST at any time t_* during the model integration, the error principal components a_n are not since the observed SST is not known at time t_*. However, the model principal components $b_n(t_*)$ are known and the $a_m(t_*)$ can be linked to the $b_n(t_*)$ as best as possible using a least-squares regression relationship based on past data. Specifically, for each m,

$$F = \sum_t \left(a_m(t) - \sum_{n=1}^{8} r_{mn} b_n(t) \right)^2$$ (10.4)

is minimized by differentiating F by r_{mk} and setting the result to zero to give

$$\sum_t \left(a_m(t) - \sum_{n=1}^{8} r_{mn} b_n(t) \right) b_k(t) = 0.$$ (10.5)

The principal components are uncorrelated in time (see Section A.2.2 of Appendix A) so

$$\sum_t b_n(t) b_k(t) = 0 \quad n \neq k.$$ (10.6)

Use of this result in (10.5) gives

$$r_{mk} = \sum_t a_m b_k \Big/ \sum_t b_k^2.$$ (10.7)

The above calculations were all done for the 16-year training period 1970–1985. As pointed out by Barnett et al. (1993), the same SST anomaly can produce a different anomalous wind field depending on the calendar month of the year. This is because the anomalous atmospheric heating depends on the total SST (Section 6.5) and so for the same SST anomaly, the total SST can be different depending on the average SST for the particular calendar month. To take this into account, the above calculations were done separately for each calendar month.

During the running of the coupled model, the bias correction for the SST at a new time t_* is carried out as follows. At the time t_* all the model data are known so, in the notation of Appendix A, the model data vector $\mathbf{X}(t_*)$ is also known. Since the EOF pattern vectors $\mathbf{p}^{(n)}$ are known from the training period, the model principal components at time t_* can be found from (see (A.10))

$$b_n(t_*) = \mathbf{p}^{(n)} \cdot \mathbf{X}(t_*). \tag{10.8}$$

Having obtained the $b_n(t_*)$, we can estimate $a_m(t_*)$ from the regression relationship (see (10.4))

$$a_m(t_*) \approx \sum_{n=1}^{8} r_{mn} b_n(t_*) \tag{10.9}$$

and hence $E(x, y, t_*)$ from (10.3) since the $e_n(x, y)$ are known. The model bias-corrected SST should be as close as possible at time t_* to the smoothed observations, so based on (10.2),

$$\text{SST}_{\text{corr}}(x, y, t_*) = \text{SST}_{\text{mod}}(x, y, t_*) - E(x, y, t_*). \tag{10.10}$$

The other two physical variables, anomalous wind stress and anomalous sea level, are corrected in a similar way to SST. The only difference is that each variable uses eight EOF modes for the observed fields and in (10.3).

The above model bias correction procedure essentially adds an interactive statistical component to the dynamical model since at every time step it corrects the model physics. It also decreases the 'shock' of assimilating observational data into the model since the corrected model variables are now much closer to the data being assimilated. Coupled model predictions are considerably improved (Chen et al. 2000).

10.2.3. Nudging

A simple way to assimilate observational data into a coupled model is by 'nudging' (see, e.g., Chen et al. 1995, 1997). For example, Chen et al. (1995, 1997) nudged the observed anomalous wind stress τ_{obs} into the model at each time step up to forecast time by modifying the model anomalous wind stress τ_{mod} to $\alpha \tau_{\text{obs}} + (1 - \alpha) \tau_{\text{mod}}$ where α is a function of latitude. If $\alpha = 1$ everywhere, no account is taken of the coupled model dynamics during the initialization and the coupled model system may not be well balanced when the forecast begins. The resultant 'shock' may greatly reduce the model's ability to forecast. On the other hand, if $\alpha = 0$ the coupled model has no knowledge of the initial conditions. The rule of thumb (Chen et al. 1997) is to weight the model winds as much as possible where they are close to the observations. Therefore for the Lamont model α ($= 0.25$) is smallest near the equator, where the model winds are most reliable. A similar nudging scheme was used later to assimilate anomalous sea level into the Lamont model (Chen et al. 1998).

10.3. THE KALMAN FILTER

The basic idea of the Kalman filter is to combine the model forecast field $\mathbf{w}_f$ with the observations $\mathbf{w}_0$ to produce an analysis field $\mathbf{w}_a$ that is as close as possible to the true state $\mathbf{w}$. In other words, we wish to minimize the variance of the analysis error, the scalar

$$G = \langle (\mathbf{w} - \mathbf{w}_a)^T (\mathbf{w} - \mathbf{w}_a) \rangle \qquad (10.11)$$

where T denotes the transpose and the angle brackets denote the expected value of the ensemble. Our analysis field is a linear combination of the model forecast field and an improvement proportional to the discrepancy between the model forecast field $\mathbf{w}_f$ and the observations $\mathbf{w}_0$. Generally $\mathbf{w}_f$ has a large number M of components because it has values at many grid points while the $m \times 1$ vector $\mathbf{w}_0$ has a much smaller number m of components. Thus the difference between the observed and model fields is of the form $\mathbf{w}_0 - H\mathbf{w}_f$, the appropriate $m \times M$ matrix H multiplying $\mathbf{w}_f$ so that each element of $H\mathbf{w}_f$ corresponds to the same spatial location as the corresponding observation. The analysis field $\mathbf{w}_a$ is therefore of the form

$$\mathbf{w}_a = \mathbf{w}_f + K(\mathbf{w}_0 - H\mathbf{w}_f), \qquad (10.12)$$

the $M \times m$ matrix K being chosen so that G in (10.11) is minimized.

Since $\mathbf{w}$ is the true state vector, we may write that the observed state

$$\mathbf{w}_0 = H\mathbf{w} + \mathbf{e} \qquad (10.13)$$

where $\mathbf{e}$ is the observational error. Thus from (10.12) and (10.13) the analysis field

$$\mathbf{w}_a = \mathbf{w}_f + K(H(\mathbf{w} - \mathbf{w}_f) + \mathbf{e}). \qquad (10.14)$$

Hence in (10.11)

$$G = \langle (\mathbf{y} - K\mathbf{x})^{\mathrm{T}}(\mathbf{y} - K\mathbf{x}) \rangle \qquad (10.15)$$

where, in this context,

$$\mathbf{y} = \mathbf{w} - \mathbf{w}_{\mathrm{f}} \qquad (10.16)$$

and

$$\mathbf{x} = H(\mathbf{w} - \mathbf{w}_{\mathrm{f}}) + \mathbf{e}. \qquad (10.17)$$

The matrix K that minimizes (10.15) is (see Appendix 10.A at the end of this chapter)

$$K = \langle \mathbf{y}\mathbf{x}^{\mathrm{T}} \rangle \langle \mathbf{x}\mathbf{x}^{\mathrm{T}} \rangle^{-1}. \qquad (10.18)$$

Substituting (10.16) and (10.17) into (10.18) and noting that $\mathbf{e}$ is not correlated with $\mathbf{w} - \mathbf{w}_{\mathrm{f}}$ gives

$$K = P_{\mathrm{f}} H^{\mathrm{T}}(HP_{\mathrm{f}}H^{\mathrm{T}} + E_0)^{-1} \qquad (10.19)$$

where P_{f}, the model error covariance, is given by

$$P_{\mathrm{f}} = \langle (\mathbf{w} - \mathbf{w}_{\mathrm{f}})(\mathbf{w} - \mathbf{w}_{\mathrm{f}})^{\mathrm{T}} \rangle \qquad (10.20)$$

and E_0, the observational error covariance, is

$$E_0 = \langle \mathbf{e}\mathbf{e}^{\mathrm{T}} \rangle. \qquad (10.21)$$

Note that (10.19) does what you expect; when the observational error covariance becomes large so that E_0 dominates $(HP_{\mathrm{f}}H^{\mathrm{T}} + E_0)$ then from (10.19) $K \to 0$ and, by (10.14) $\mathbf{w}_{\mathrm{a}} \to \mathbf{w}_{\mathrm{f}}$, i.e., the model estimate is best when the observational error is large.

From (10.12), in order to calculate the desired analysis field $\mathbf{w}_{\mathrm{a}}$ we need K, $\mathbf{w}_{\mathrm{f}}$ and $\mathbf{w}_0$. The latter two vectors are given but, by (10.19), K can only be calculated if P_{f} and E_0 are known. The optimal interpolation technique, which has been used to assimilate data into models by using the analysis field $\mathbf{w}_{\mathrm{a}}$ to replace $\mathbf{w}_{\mathrm{f}}$, specifies both the observational error covariance E_0 and a time-invariant model error covariance P_{f}. In the Kalman filter P_{f} is not specified but varies from one time step to the next as the model evolves.

To calculate P_f for the Kalman filter at time t begin from the forecast equation

$$\mathbf{w}_f(t) = A\mathbf{w}_a(t-1) + B\boldsymbol{\tau}(t-1) \tag{10.22}$$

where all quantities are known at time $t = 0$, $\boldsymbol{\tau}$ is the wind stress and A and B are known matrices from the dynamics. Analogous to (10.22), the true state vector $\mathbf{w}(t)$ at time t can be written as

$$\mathbf{w}(t) = A\mathbf{w}(t-1) + B\boldsymbol{\tau}(t-1) + \boldsymbol{\varepsilon}(t-1) \tag{10.23}$$

where the 'system' error $\boldsymbol{\varepsilon}$ takes into account the model and wind errors in going from $t-1$ to t. Substitution of (10.22) and (10.23) into (10.20) and assuming that the error $\boldsymbol{\varepsilon}$ is not correlated with $\mathbf{w} - \mathbf{w}_f$ gives

$$P_f(t) = A\langle(\mathbf{w}(t-1) - \mathbf{w}_a(t-1))(\mathbf{w}(t-1) - \mathbf{w}_a(t-1))^T\rangle A^T + \langle\boldsymbol{\varepsilon}\boldsymbol{\varepsilon}^T\rangle \tag{10.24}$$

where $\langle\boldsymbol{\varepsilon}\boldsymbol{\varepsilon}^T\rangle$ is the $M \times M$ system error covariance matrix. Equation (10.24) may be written in the more compact form

$$P_f(t) = AP_a(t-1)A^T + \langle\boldsymbol{\varepsilon}\boldsymbol{\varepsilon}^T\rangle \tag{10.25}$$

where the analysis error covariance matrix at time $t-1$ is

$$P_a(t-1) = \langle(\mathbf{w}(t-1) - \mathbf{w}_a(t-1))(\mathbf{w}(t-1) - \mathbf{w}_a(t-1))^T\rangle. \tag{10.26}$$

This matrix can be calculated from (see Appendix 10.B at the end of this chapter)

$$P_a = (I - KH)P_f \tag{10.27}$$

with I as the identity matrix and P_a and P_f both at time $t-1$.

Thus the Kalman filter estimate of the analysis field $\mathbf{w}_a$ is given by (10.12) with K as in (10.19). In (10.19) the observational error covariance matrix E_0 must be specified but P_f is calculated from (10.25) and (10.27). To do this calculation, however, the system error variance matrix $\langle\boldsymbol{\varepsilon}\boldsymbol{\varepsilon}^T\rangle$ must be specified.

Practical implementation of Kalman filter theory to the ENSO problem has its difficulties (Cane et al. 1996). One is that the structures of the errors in the observations and the model are not well known. Another difficulty is computational cost; if M is $O(10^5)$ then P_f has $O(10^{10})$ elements and updating $P_f(t-1)$ is often beyond our computational capabilities. But in all this Cane et al. recognized that since the observations are widely scattered and since the model is only accurate on the large scale, at best only large-scale variability can be determined. Cane et al. implemented Kalman filter theory by restricting the error structure to the large scale and by efficiently representing the data by a comparatively small truncated number of EOFs.

10.4. ADJOINT DATA ASSIMILATION

Suppose a model is integrated from some initial state to obtain model solutions over a data assimilation period, a period of time when observations are available. Adjoint data assimilation can be used to choose the initial conditions so that the data are assimilated into the model as well as possible in the sense that, over the assimilation period, the departure of the model from the observations is as small as possible. A simple example illustrating the technique is presented below.

Let the model variables be $\mathbf{q}(t)$ where the column vector $\mathbf{q}(t)$ at month t has M components. For example, $\mathbf{q}$ could represent two different physical quantities at $M/2$ spatial grid points. We have available observations $\mathbf{r}(t)$, $t = 1, 2, \ldots, N$. We want to choose model initial conditions $\mathbf{q}(0)$ so that the cost function

$$J = \sum_{t=1}^{N} \sum_{j=1}^{M} (q_j(t) - r_j(t))^2 / (\sigma_j)^2 \qquad (10.28)$$

is as small as possible. In (10.28), σ_j is the (given) spatially varying estimated error in the observed values of r_j. By weighting $(q_j - r_j)^2$ by σ_j^{-2} in the formulation of the cost function J, the difference between the model and observations appropriately means less to the cost function when the error is large.

The dynamical equations governing the ocean model can be written in the form

$$\mathbf{q}(t) = A\mathbf{q}(t-1) + B\mathbf{X}(t-1) \qquad (10.29)$$

where A and B are $M \times M$ matrices known from the dynamics and $\mathbf{X}(t-1)$ can be determined from the given wind stress forcing at time $t-1$. Given $\mathbf{q}(0)$, (10.29) can be used to determine $\mathbf{q}$ at all times t and hence the cost function J in (10.28) can be calculated. Our problem is to find $\mathbf{q}(0)$ so that J is minimized subject to the model equation constraints (10.29).

We can solve such a constrained problem using the M dimensional column vector of Lagrange multipliers $\boldsymbol{\lambda}(t)$. We take the dot product of $\boldsymbol{\lambda}(t)$ with $\mathbf{q}(t) - A\mathbf{q}(t-1) - B\mathbf{X}(t-1)$ for $t = 1, 2, \ldots, N$, sum the result over these times t and add to J to obtain

$$L = J + \sum_{t_*=1}^{N} \sum_{i=1}^{M} \lambda_i(t_*) \left(q_i(t_*) - \sum_{j=1}^{M} A_{ij} q_j(t_* - 1) - (B\mathbf{X}(t_* - 1))_i \right) \quad (10.30)$$

where A_{ij} is the i, jth element of the matrix A and $(B\mathbf{X}(t_* - 1))_i$ is the ith element of the vector $B\mathbf{X}(t_* - 1)$. At minimum L

$$\frac{\partial L}{\partial \lambda_k(t)} = 0 \qquad t = 1, \ldots, N; \quad k = 1, 2, \ldots, M, \qquad (10.31a)$$

$$\frac{\partial L}{\partial q_k(t)} = 0 \qquad t = 1, \ldots, N; \quad k = 1, 2, \ldots, M, \qquad (10.31b)$$

and

$$\frac{\partial L}{\partial q_k(0)} = 0 \quad k = 1, 2, \ldots, M. \qquad (10.31c)$$

The first set of conditions (10.31a) implies that the model equations (10.29) are satisfied at all grid points for $t = 1, 2, \ldots, N$ as required. The second set of conditions implies, from (10.30) that, for $k = 1, \ldots, M$ and $t = 1, \ldots, N$

$$\frac{\partial J}{\partial q_k(t)} + \lambda_k(t) - \sum_{i=1}^{M} \lambda_i(t+1) A_{ik} = 0. \qquad (10.32)$$

The third set of conditions (10.31c) gives, using (10.32),

$$\frac{\partial J}{\partial q_k(0)} = \sum_{i=1}^{M} \lambda_i(1) A_{ik} \quad k = 1, 2, \ldots, M. \qquad (10.33)$$

Note that

$$\sum_{i=1}^{M} \lambda_i(t+1) A_{ik} = (A^{\mathrm{T}} \boldsymbol{\lambda}(t+1))_k. \qquad (10.34)$$

Equations (10.32) and (10.33), which involve the transpose of A (see (10.34)), are called the adjoint equations.

Equations (10.29), (10.32) and (10.33) can be used to find $\mathbf{q}(0)$ so that J is minimized. Given a first guess $\mathbf{q}(0)$, the model can be integrated using (10.29) to find $\mathbf{q}(t)$ for $t = 1, \ldots, N$. Hence $\partial J / \partial q_k(t)$ can be found from (10.28) for $k = 1, \ldots, M$ and $t = 1, \ldots, N$. Since $\partial J / \partial q_k(t)$ is known, under the condition $\boldsymbol{\lambda}(N+1) = \mathbf{0}$, the adjoint equations (10.32), written as

$$\lambda_k(t) = \sum_{i=1}^{M} \lambda_i(t+1) A_{ik} - \frac{\partial J}{\partial q_k(t)}, \qquad (10.35)$$

can be used to find $\lambda_k(t)$ for all $t = 1, \ldots, N$ by starting at $t = N$ and going back in time. Specifically, $\lambda_k(N)$ can be found from (10.35) for $t = N$, then knowledge of $\lambda_k(N)$ enables $\lambda_k(N-1)$ to be found from the $t = N - 1$ equations and so on. With $\boldsymbol{\lambda}(1)$ known, $\partial J / \partial q_k(0)$ can be calculated from (10.33). Now that we have $\partial J / \partial q_k(0)$, we know how the cost function J changes locally with the initial conditions. Therefore we may use it and our first guess of $\mathbf{q}(0)$ to obtain an improved estimate of $\mathbf{q}(0)$ that decreases J. This can be done using a conjugate gradient method (see, e.g., Moore 1991). The process is then repeated until minimum J and the corresponding $\mathbf{q}(0)$ are determined.

The above example illustrates the basic concept of adjoint data assimilation but few models are this simple. One generalization is that the cost function J is usually a more complicated function of the model variables. For example, Kleeman et al. (1995) include an extra term in the cost function so that the model zonal scales are as smooth as the analyzed observations. Often the observed data will be fewer than those of the model and this will also affect J. Another feature particular to the above model is that the model equations are a strong constraint in that they are satisfied exactly. In many cases the assimilation procedure will recognize that the model itself is imperfect and the model equations are weakly constrained, i.e., they are only satisfied to within some specified error (see, e.g., Bennett et al. 1998).

10.5. Dynamical Model Forecast Performance

Many dynamical models operationally predict ENSO variability monthly by predicting monthly values of NINO3.4 for various lead times. We will consider the performance of six of these (see Table 10.1). These models were chosen based on their performance data in the published literature and, in some cases, supplemental data provided by the authors. There are three coupled ocean–atmosphere general circulation models, an

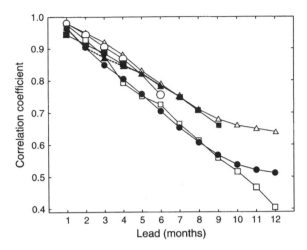

Fig. 10.1 Correlation of 'predictions' of NINO3.4 with observed NINO3.4 by each of the six models in Table 10.1 for the periods indicated. The solid circles represent the Lamont model (LDEO5), the solid squares the NCEP (CFS) model, the solid triangles and dashed lines the ECHAM-MOM model, the open circles the ECMWF (S2) model, the open squares the COLA model and the open triangles the Korean hybrid model. In most cases the 'predictions' were hindcasts.

anomaly coupled general circulation model and two coupled intermediate models, one of which uses a statistically based rather than dynamical atmospheric model. Model details can be found in the references in Table 10.1.

Figure 10.1 documents the models' hindcast performance. In the figure a 1-month lead means that the (say) April value of NINO3.4 is predicted from data up to the end of March; a 2-month lead time would correspond to the prediction of the May NINO3.4 value, etc. An exception is Korean hybrid coupled model which predicts the El Niño index NINO3 (the sea surface temperature anomaly averaged over the eastern equatorial Pacific region 5°S–5°N, 150°W–90°W) instead of NINO3.4. Model hindcasts are impressive and similar; all have correlations greater than 0.7 at a lead time of 6 months and one has a correlation greater than 0.6 at a lead time of a year.

Many of the model hindcasts have used data from the hindcast period to determine some model parameters. In such cases one might suspect that the hindcast skill might be artificially high. But the Lamont model LDEO5, using only SST data during 1976–1995 for training the model, hindcasts NINO3.4 nearly as well for 128 years of independent data (Fig. 10.2). These hindcasts, as far back as the mid-nineteenth century, suggest that long-range forecasts can be made successfully even though global warming and decadal changes are modulating the 'mean' state of the ocean and atmosphere.

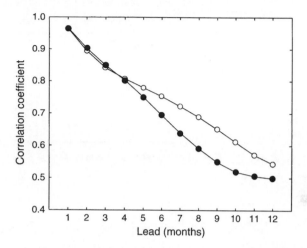

Fig. 10.2 Correlation of model LDEO5 NINO3.4 with observed NINO3.4 as a function of forecast lead for the training period 1976–1995 (open circles) and the independent prediction period 1856–1975 and 1996–2003 (closed circles).

APPENDIX 10.A. MINIMIZATION OF (10.15)

Written out explicitly, (10.15) is

$$G = \left\langle \sum_{i=1}^{M} \left(y_i - \sum_{j=1}^{m} K_{ij} x_j \right)^2 \right\rangle \tag{10A.1}$$

where y_i is the ith element of **y**, x_j the jth element of **x** and K_{ij} the ijth element of K. A necessary condition for the minimization of F is that

$$\frac{\partial G}{\partial K_{pq}} = 0 \qquad p = 1, \ldots, M; \quad q = 1, \ldots, m. \tag{10A.2}$$

This yields

$$\left\langle 2 \left(y_p - \sum_{j=1}^{m} K_{pj} x_j \right) (-x_q) \right\rangle = 0,$$

i.e.,

$$\langle y_p x_q \rangle = \sum_{j=1}^{m} K_{pj} \langle x_j x_q \rangle. \tag{10A.3}$$

In matrix form (10A.3) is

$$\langle \mathbf{y}\mathbf{x}^{\mathrm{T}}\rangle = K\langle \mathbf{x}\mathbf{x}^{\mathrm{T}}\rangle \tag{10A.4}$$

from which we obtain (10.18) of the main text.

APPENDIX 10.B. CONNECTING P_a AND P_f

From (10.14)

$$\mathbf{w} - \mathbf{w}_a = (I - KH)(\mathbf{w} - \mathbf{w}_f) - K\mathbf{e}. \tag{10B.1}$$

Substitution into (10.26) of the main text and using the result that the error $\mathbf{e}$ is uncorrelated with $\mathbf{w}$ or $\mathbf{w}_f$ gives

$$P_a = (I - KH)\langle (\mathbf{w} - \mathbf{w}_f)(\mathbf{w} - \mathbf{w}_f)^{\mathrm{T}}\rangle (I - KH)^{\mathrm{T}} + K\langle \mathbf{e}\mathbf{e}^{\mathrm{T}}\rangle K^{\mathrm{T}} \tag{10B.2}$$

or, by (10.20) and (10.21),

$$P_a = (I - KH)P_f\left(I - H^{\mathrm{T}}K^{\mathrm{T}}\right) + KE_0 K^{\mathrm{T}}. \tag{10B.3}$$

This may also be written as

$$P_a = (I - KH)P_f - (I - KH)P_f H^{\mathrm{T}}K^{\mathrm{T}} + KE_0 K^{\mathrm{T}}. \tag{10B.4}$$

But upon multiplying (10.19) by $(HP_f H^{\mathrm{T}} + E_0)$ and then the result by K^{T} we have

$$K\left(HP_f H^{\mathrm{T}} + E_0\right)K^{\mathrm{T}} = P_f H^{\mathrm{T}}K^{\mathrm{T}}$$

or, equivalently,

$$KE_0 K^{\mathrm{T}} = (I - KH)P_f H^{\mathrm{T}}K^{\mathrm{T}}. \tag{10B.5}$$

Substitution of (10B.5) into (10B.4) then gives the required result (10.27) of the main text.

> ## REFERENCES

Barnett, T. P., M. Latif, N. Graham, M. Flugel, S. Pazan, and W. White, 1993: ENSO and ENSO-related predictability. Part I: Prediction of equatorial Pacific sea surface temperature with a hybrid coupled ocean–atmosphere model. *J. Clim.*, **6**, 1545–1566.

Bennett, A. F., B. S. Chua, D. E. Harrison, and M. J. McPhaden, 1998: Generalized inversion of tropical atmosphere–ocean data and a coupled model of the tropical Pacific. *J. Clim.*, **11**, 1768–1792.

Cane, M. A., A. Kaplan, R. N. Miller, B. Tang, E. C. Hackert, and A. J. Busalacchi, 1996: Mapping tropical Pacific sea level: Data assimilation via a reduced state space Kalman filter. *J. Geophys. Res.*, **101**, 22,599–22,617.

Chen, D., M. A. Cane, A. Kaplan, S. E. Zebiak, and D. Huang, 2004: Predictability of El Niño over the past 148 years. *Nature*, **428**, 733–736.

Chen, D., M. A. Cane, S. E. Zebiak, R. Cañizares, and A. Kaplan, 2000: Bias correction of an ocean–atmosphere coupled model. *Geophys. Res. Lett.*, **27**(16), 2585–2588.

Chen, D., M. A. Cane, S. E. Zebiak, and A. Kaplan, 1998: The impact of sea level data assimilation on the Lamont model prediction of the 1997/98 El Niño. *Geophys. Res. Lett.*, **25**, 2837–2840.

Chen, D., S. E. Zebiak, A. J. Busalacchi, and M. A. Cane, 1995: An improved procedure for El Niño forecasting: Implication for predictability. *Science*, **269**, 1699–1702.

Chen, D., S. E. Zebiak, M. A. Cane, and A. J. Busalacchi, 1997: Initialization and predictability of a coupled ENSO forecast model. *Mon. Weather Rev.*, **125**, 773–788.

DeWitt, D. G., 2005: Retrospective forecasts of interannual sea surface temperature anomalies from 1982 to present using a directly coupled atmosphere–ocean general circulation model. *Mon. Weather Rev.*, **133**, 2972–2995.

Kang, I. S., and J. S. Kug, 2000: An El Niño prediction system using an intermediate ocean and a statistical atmosphere. *Geophys. Res. Lett.*, **27**(8), 1167–1170.

Kirtman, B. P., 2003: The COLA anomaly coupled model: Ensemble predictions. *Mon. Weather Rev.*, **131**, 2324–2341.

Kleeman, R., A. M. Moore, and N. R. Smith, 1995: Assimilation of subsurface thermal data into an intermediate tropical coupled ocean–atmosphere model. *Mon. Weather Rev.*, **123**, 3103–3113.

Kug, J.-S., I.-S. Kang, and S. E. Zebiak, 2001: The impacts of the model assimilated wind stress data in the initialization of an intermediate ocean and the ENSO predictability. *Geophys. Res. Lett.*, **28**(19), 3713–3716.

Moore, A. M., 1991: Data assimilation in a quasi-geostrophic open-ocean model of the Gulf Stream region using the adjoint method. *J. Phys. Oceanogr.*, **21**, 398–427.

Pierce, D. W., 1996: *The hybrid coupled model, version 3: Technical notes.* SIO Reference Series No. 96-27, Scripps Institution of Oceanography, University of California, San Diego, August, 1996.

Saha, S., S. Nadiga, C. Thiaw, J. Wang, W. Wang, Q. Zhang, H. M. Van den Dool, H. L. Pan, S. Moorthi, D. Behringer, D. Stokes, M. Peña, S. Lord, G. White, W. Ebisuzaki, P. Peng, and P. Xie, 2006: The NCEP climate forecast system. *J. Clim*, **19**, 3483–3517.

Van Oldenborgh, G. J., M. A. Balmaseda, L. Ferranti, T. N. Stockdale, and D. L. T. Anderson, 2005: Did the ECMWF seasonal forecast model outperform statistical ENSO forecast models over the last 15 years? *J. Clim.*, **18**, 3240–3249.

Zebiak, S. E., and M. A. Cane, 1987: A model El Niño–Southern Oscillation. *Mon. Weather Rev.*, **115**, 2262–2278.

Zhang, R. H., S. E. Zebiak, R. Kleeman, and N. Keenlyside, 2003: A new intermediate coupled model for El Niño simulation and prediction. *Geophys. Res. Lett.*, **30**(19), Art. No. 2012, doi: 10.1029/2003GL018010.

CHAPTER ELEVEN

ENSO FORECASTING USING STATISTICAL MODELS

Contents

11.1. OVERVIEW

As noted in Chapter 8, major ENSO indices such as the El Niño index NINO3.4 and the Tahiti minus Darwin Southern Oscillation Index tend to have the same sign over the latter half of the calendar year. During this time of the year, a good long-range forecast is the persistence forecast. For example, at the end of July, it is possible to forecast the January value of NINO3.4 from the July value with a correlation of 0.85. This remarkable ENSO property is associated with a growing, eastward propagating coupled ocean–atmosphere ENSO disturbance in the western and central equatorial Pacific (see Chapter 8).

Persistence is, however, the crudest statistical forecast model available. It fails badly during the first half of the calendar year when we try to predict across the Northern Hemisphere Spring. For example, while the correlation from the July NINO3.4 value to January NINO3.4 the following year is 0.85, the correlation from January to July across the spring is only 0.03.

In this chapter, we will discuss better statistical methods to predict ENSO. Dynamical prediction models were discussed in Chapter 10.

Many of the statistical ENSO prediction models have been summarized by Barnston et al. (1994) and Barnston et al. (1999). In this text, I will not attempt to describe all of these models but rather select a few that illustrate some of the statistical techniques used for ENSO prediction. The remaining sections of this chapter describe these statistical models, beginning with the simplest.

11.2. THE CLIMATOLOGY AND PERSISTENCE FORECASTING SCHEME

In January 1997 the El Niño index NINO3.4 was $-0.4°C$ but then it began to rise to $-0.3°C$ in February and $0.0°C$ in March. Since this El Niño index and others frequently change sign in the boreal spring, and since NINO3.4 is dominated by interannual frequencies, it was much more likely that NINO3.4 would continue to rise and an El Niño would develop than that it would persist at $0°C$. A prediction scheme that takes advantage of both trends at certain times of the calendar year and persistence at other times is the Climatology and Persistence (CLIPER) forecasting scheme of Knaff and Landsea (1997).

The CLIPER scheme predicts an ENSO time series (e.g., NINO3.4) using a linear sum of ENSO indices and/or their trends. For example, suppose it is January 1, and we want to predict the April–May–June (AMJ) average of NINO3.4. The CLIPER prediction is (Knaff and Landsea 1997)

$$\text{NINO3.4(AMJ)} = 0.121 + 0.492(\text{NINO3.4(December)}$$
$$- \text{NINO3.4(October)})$$
$$+ 0.864(\text{NINO4(OND)} - \text{NINO4(JAS)}) \quad (11.1)$$

where NINO4 is the SST anomaly averaged over 5°S–5°N, 150°W–160°E and NINO4 (OND) and NINO4 (JAS) refer, respectively, to the most recent October–December and July–September 3-month averages of that El Niño index. The constants 0.121, 0.492 and 0.864 in (11.1) were determined by least squares fitting the time series in (11.1) using monthly data over the period 1950–1994. In theory, since all the time series are for anomalies, the constant term (0.121) should actually be zero. But the anomalies were calculated from a 1950–1979 climatology rather than the 1950–1994 training period, so the small constant term is present.

Predictions from different times of the year for different lead times yield other linear combinations of possible predictors of NINO3.4. For each starting time and each lead, Knaff and Landsea considered 14 possible predictors and used objective criteria to limit the total number used to four or less. The possible predictors included most recent averages and trends of NINO3.4, NINO4, the Tahiti minus Darwin SOI, NINO3 (average SST anomaly for the region 5°S–5°N, 150°W to 90°W) and NINO1+2 (average SST anomaly for the region 0°–10°S, 90°W to 80°W).

A danger in considering so many possible predictors is that 'over-fitting' will occur, i.e., the resultant regression equation will provide excellent hindcasts for the training period but poor forecasts for independent data. The CLIPER scheme has been 'frozen' since February 1993, so independent forecasts have been made since then. Figure 11.1 shows the correlation of NINO3.4 predicted by CLIPER against observed NINO3.4 at leads of 1–12 months. The CLIPER scheme makes 3-month average forecasts rather than monthly forecasts, and we define a (say) forecast of

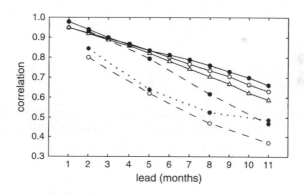

Fig. 11.1 Correlation performance as a function of forecast lead for predicting the El Niño index NINO3.4: open circles, dashed line, the canonical correlation analysis model of Barnston and Ropelewski (1992) for the period 1957–1990; closed circles, dotted line, the linear inverse model of Penland and Magorian (1993) and Penland and Sardeshmukh (1995) for the period 1970–1993; open circles, solid line, the constructed analogue (CA) model of van den Dool (1994a, b) for the period 1981–2001; open triangles, solid line, the CA model for the period 1955–2001; closed circles, dashed line, the climatology and persistence (CLIPER) model of Knaff and Landsea (1997) for the period January 1993 to November 2006; closed circles, solid line, the Clarke and Van Gorder model using (11.3) for the period 1981–2001. All correlation results are cross-verified hindcasts except for the Knaff and Landsea (1997) CLIPER model for which the correlation results are from operational forecasts. Two time intervals are plotted for the CA model, one over the long record 1955–2001 and the other over the more recent period 1981–2001 when the data were more accurate and complete and there were two large El Niños. There is little difference in the lead correlations.

April–May–June NINO3.4, given data up to the end of December as a 5-month lead forecast. Knaff and Landsea (1997) argue that the CLIPER scheme should be a baseline for testing the skill of ENSO prediction models since it involves no physics, just persistence, trends and the climatology of the predictors. The CLIPER scheme sets a high bar for model performance; over nearly 14 years of actual forecasts, the CLIPER scheme predicts NINO3.4 with a correlation of better than 0.6, eight months in advance (see Fig. 11.1).

11.3. A Precursor ENSO Prediction Model

As we saw in Section 7.3.3, there are physical arguments suggesting that the thermocline depth anomaly in the central equatorial Pacific, D_{cen}, should be approximately proportional to $(\partial/\partial t)(T_{cen})$, where T_{cen} is the SSTA in the central equatorial Pacific. But as noted in Section 7.3.2, T_{cen} is approximately proportional to the El Niño index NINO3.4 and D_{cen} is approximately proportional to $\int_S D\,dS$ and hence to $\overline{D}$, the anomaly of the 20°C isotherm depth averaged zonally across the Pacific from 5°S to 5°N. Based on these results and (7.33),

$$\frac{\partial}{\partial t}(\text{NINO3.4}) \approx b\overline{D} \qquad (11.2)$$

for some positive constant b. Equation (11.2) suggests that a linear combination of $\overline{D}$ and NINO3.4 should be a powerful ENSO predictor; NINO3.4 tells us the present value of NINO3.4 and $b\overline{D}$ tells us the direction NINO3.4 is headed.

If NINO3.4 and $\overline{D}$ were sinusoidal, (11.2) would imply that $\overline{D}$ would lead NINO3.4 by a quarter of a period. This result suggests that for the irregular interannual oscillations characteristic of NINO3.4 and $\overline{D}$, $\overline{D}$ may lead NINO3.4 by several months. We can see this in Fig. 11.2; typically, $\overline{D}$ leads NINO3.4 by about 5 months. Figure 11.3 shows that $\overline{D}$ is particularly useful for jumping the spring persistence barrier mentioned in Chapter 8 and in the overview of this chapter. For example, the correlation of the December, January, February and March values of $\overline{D}$ with NINO3.4 in July and for several months following is almost always greater than 0.6 and are typically about 0.7.

Another useful precursor that can jump the spring persistence barrier is associated with the zonal equatorial wind stress. As noted in Section 8.4, zonal equatorial wind stress anomalies can usually be seen in the eastern Indian Ocean and far western equatorial Pacific *prior to* the boreal spring.

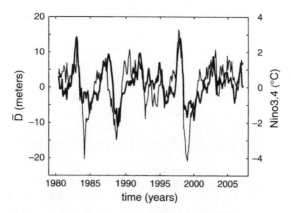

Fig. 11.2 NINO3.4 (thin line) and $\overline{D}$, the 20°C isotherm depth anomaly averaged zonally across the Pacific from 5°S to 5°N (thick line). NINO3.4 and $\overline{D}$ are maximally correlated ($r = 0.65$) when $\overline{D}$ leads NINO3.4 by 5 months. In the plot, $\overline{D}$ has been shifted forward by 5 months so that the maximum correlation can be seen more clearly.

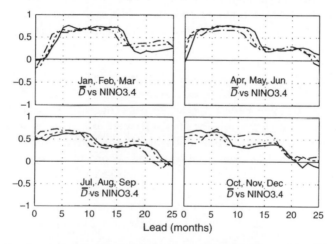

Fig. 11.3 Lead correlations of the anomalous equatorial heat content $\overline{D}(t)$ for each calendar month with NINO3.4 ($t +$ lead) for the period January 1980–December 2001. A positive lead corresponds to $\overline{D}$ leading NINO3.4. The 20°C isotherm depth is used to estimate $\overline{D}(t)$. The first month in each quarter is solid, the second is dashed and the third is dot-dashed. (Adapted from Clarke and Van Gorder 2003.)

The anomalies appear to slowly propagate eastward from the equatorial eastern Indian Ocean to the equatorial Pacific (see Fig. 8.10). A monthly wind predictor index τ, averaged in space and time along the propagation path to maximize the signal (see Table 8.1), is positively correlated with NINO3.4 (see Fig. 8.11). Correlations of τ in December, January, February and March with NINO3.4 in the following July and later months are about 0.6.

Based on the precursor properties of NINO3.4, τ and $\overline{D}$, Clarke and Van Gorder (2003) constructed the simple precursor regression prediction model. If $t = 0$ is the present time so that NINO3.4(0), $\tau(0)$ and $\overline{D}(0)$ are current values of the three precursors, then NINO3.4(Δt), the predicted value of NINO3.4 for a lead time Δt, is given by

$$\text{NINO3.4}(\Delta t) = \alpha_1 \text{NINO3.4}(0) + \alpha_2 \tau(0) + \alpha_3 \overline{D}(0). \qquad (11.3)$$

The coefficients α_1, α_2 and α_3 depend on Δt and calendar month because of the phase-locking of the precursors to the calendar year. For example, when $t = 0$ corresponds to July and $\Delta t = 1, 2, \ldots, 7$ months, then $\alpha_1 \text{NINO3.4}(0)$ dominates (11.3) because NINO3.4(0) is very highly correlated with NINO3.4 later in the year and the beginning of the next year. On the other hand, if $t = 0$ corresponds to (say) January and prediction across the spring persistence barrier is needed, ($\Delta t \geq 4$ months), then the second and third terms in (11.3) dominate. For each lead time Δt, Clarke and Van Gorder (2003) determined the $12\alpha_1$, $12\alpha_2$ and $12\alpha_3$ coefficients (one for each calendar month) by least squares fitting the right-hand side of (11.3) over a training period.

The model was cross-validated for the period January 1981 to December 2001. A given month was 'predicted' by training the model over the whole record except for a 2-year period centered on the month being predicted. Anomalies for this 'prediction' were based on seasonal cycles estimated using the data in 1981–2001 minus the 'unknown' 2 years. This process was repeated for each month in the record to obtain 'predictions' for about 20 years. As might be expected based on the precursors, the cross-validated results show that the model can predict across the persistence barrier (Clarke and Van Gorder 2003). Overall, the model exhibits good skill at lead times of at least 11 months (see Fig. 11.1).

11.4. PREDICTION USING CANONICAL CORRELATION ANALYSIS

Canonical correlation analysis (CCA) is a technique used to study the correlation structure of two random vectors $\mathbf{X}(t)$ and $\mathbf{Y}(t)$. I describe the technique in Appendix B and summarize the basic idea and how CCA can be used as a prediction method in Section 11.4.1. Then in Section 11.4.2, we will discuss how CCA has been used to predict ENSO.

11.4.1. The Basic Idea

The canonical correlation method finds a linear combination of the m predictor time series that maximizes correlation with a linear combination of the n predictand time series. Mathematically, let the m time series be written in the form of a time-dependent column vector $\mathbf{X}(t)$ of length m and let the n time series be written as a time-dependent column vector $\mathbf{Y}(t)$ of length n. All time series have a zero mean. If $\mathbf{p}$ and $\mathbf{q}$ are time-independent vectors of lengths m and n, respectively, then

$$\alpha(t) = \mathbf{p} \cdot \mathbf{X}(t) \tag{11.4}$$

and

$$\beta(t) = \mathbf{q} \cdot \mathbf{Y}(t) \tag{11.5}$$

are linear combinations of the time series. CCA maximizes the correlation between these time series. In the process (see Appendix B), it actually finds M pairs $\mathbf{p}^{(j)}$ and $\mathbf{q}^{(j)}$ of weighting vectors and M 'canonical variates' $\alpha_j(t)$ and $\beta_j(t)$. The number M is less than or equal to the minimum of m and n. The first canonical pair $\mathbf{p}^{(1)}$ and $\mathbf{q}^{(1)}$ describes those weights that maximize the correlation between the linear combination $\mathbf{p} \cdot \mathbf{X}$ of the m time series with the linear combination $\mathbf{q} \cdot \mathbf{Y}$ of the n time series; the second canonical vector pair $\mathbf{p}^{(2)}$ and $\mathbf{q}^{(2)}$ describes those weights that maximize the correlation of $\mathbf{p} \cdot \mathbf{X}$ with $\mathbf{q} \cdot \mathbf{Y}$ *and* produces linear combinations $\mathbf{p}^{(2)} \cdot \mathbf{X}$ and $\mathbf{q}^{(2)} \cdot \mathbf{Y}$ that are uncorrelated with $\mathbf{p}^{(1)} \cdot \mathbf{X}$ and $\mathbf{q}^{(1)} \cdot \mathbf{Y}$; the third canonical pair maximizes the correlation between $\mathbf{p} \cdot \mathbf{X}$ and $\mathbf{q} \cdot \mathbf{Y}$ with linear combinations $\mathbf{p}^{(3)} \cdot \mathbf{X}$ and $\mathbf{q}^{(3)} \cdot \mathbf{Y}$ that are uncorrelated with $\mathbf{p}^{(1)} \cdot \mathbf{X}$, $\mathbf{p}^{(2)} \cdot \mathbf{X}$, $\mathbf{q}^{(1)} \cdot \mathbf{Y}$ and $\mathbf{q}^{(2)} \cdot \mathbf{Y}$ and so on for the other higher canonical pairs.

If it is time t_* now, then a prediction of the time series of $\mathbf{Y}$, a time Δt in the future, can be found from a CCA of $\mathbf{X}(t_*)$ with $\mathbf{Y}(t_* + \Delta t)$. Specifically, a prediction $\hat{Y}_i(t_* + \Delta t)$ for the ith time series of $\mathbf{Y}$ is given by the weighted sum of the M canonical variates $\mathbf{p}^{(k)} \cdot \mathbf{X}(t_*) = \alpha_k(t_*)$ as (see (B.40) and (B.41))

$$\hat{Y}_i(t_* + \Delta t) = \sum_{k=1}^{M} \gamma_{ik} \alpha_k(t_*) \tag{11.6}$$

where

$$\gamma_{ik} = \sum_t Y_i(t + \Delta t) \alpha_k(t). \tag{11.7}$$

In (11.7), the sum is over past times t for which $Y_i(t + \Delta t)$ and $\alpha_k(t)$ are known. Note here that the α_k result from a CCA of $\mathbf{X}(t)$ with $\mathbf{Y}(t + \Delta t)$ [not $\mathbf{X}(t)$ with $\mathbf{Y}(t)$].

In practical cases, many of the predictor time series in $\mathbf{X}(t)$ and also in the predictand time series $\mathbf{Y}(t + \Delta t)$ are often highly correlated. This makes it difficult to use CCA as outlined above. For example, suppose all the time series in $\mathbf{X}(t)$ were identical. Then there would be an infinite number of weighting vectors $\mathbf{p}$ and $\mathbf{q}$ giving rise to maximum correlation and the problem is ill-conditioned. This difficulty can be avoided by representing $\mathbf{X}$ and $\mathbf{Y}$ in terms of empirical orthogonal functions (EOFs). Section B.5 shows that maximizing the correlation between linear combinations of the time series in $\mathbf{X}(t)$ and $\mathbf{Y}(t + \Delta t)$ is equivalent to maximizing the correlation of a linear combination of the principal components of $\mathbf{X}(t)$ with a linear combination of the principal components of $\mathbf{Y}(t + \Delta t)$. Using principal components instead of $\mathbf{X}(t)$ and $\mathbf{Y}(t + \Delta t)$ avoids the ill-conditioning since the principal components are uncorrelated with each other (see Section A.2.2). Furthermore, although there are often a large number of m EOFs for $\mathbf{X}(t)$ and n for $\mathbf{Y}(t + \Delta t)$, often much of the variability is described by the lowest few modes, the remaining variability being noisy and not related to the signal of interest. This means that only a few EOFs are needed and the signal to noise ratio increases.

11.4.2. Application of CCA to ENSO prediction

Barnett and Preisendorfer (1987) first applied CCA to the prediction of short-term climate fluctuations. The CCA statistical model of Barnston and Ropelewski (1992) (modified slightly by Smith et al. 1995) is used routinely to forecast the El Niño index NINO3.4 and we will focus on this model.

If it is time t_* now, then the Barnston and Ropelewski (1992) model has a predictand vector $\mathbf{Y}(t_* + \Delta t)$ consisting of eight equatorial SST anomaly time series, each data point being a 3-month mean. These time series are time series of SST anomalies averaged over various equatorial regions: NINO3 (the eastern equatorial Pacific, 5°S–5°N, 150°W to 90°W); NINO4 (the west-central equatorial Pacific, 5°S–5°N, 150°W–160°E); NINO3.4 (the central equatorial Pacific, 5°S–5°N, 170°W to 120°W); NINO1+2 (the eastern equatorial Pacific near the coast south of the equator, 0°S–10°S, 80°W–100°W); the far eastern equatorial Pacific north of the equator (0°N–10°N, 80°W–100°W); the western equatorial Pacific (10°S–10°N, 140°E–180°E); and the far eastern equatorial Indian Ocean south of the equator (5°S–15°S, 90°E–110°E). The predictor time series $\mathbf{X}(t)$ are in the form of 3-month means and consist of the above eight time series and sea level pressure (SLP) time series at 712 grid points between 40°S and 70°N. At each grid point, there are four 3-month mean SLP predictor time series covering intervals 3 months, 4–6 months, 7–9 months and 10–12 months prior to forecast time. Since there are 712 grid points, there are $712 \times 4 = 2848$ SLP predictor time series. These time series greatly outnumber the eight SST time series, so Barnston and Ropelewski weighted the latter so that they had more of an influence. For reasons discussed earlier, the $\mathbf{X}$ and $\mathbf{Y}$ time series were each expressed as a sum of EOFs and each sum was truncated to a few modes (always 3 for $\mathbf{X}$ and nearly always 3 (rarely 2) for $\mathbf{Y}$). CCA predictions are made based on a CCA of the principal components (see Appendix B). Separate analysis is done for each lead time and each of 12 forecast times for each month of the calendar year. The forecasts for each time series in $\mathbf{Y}$ are for 3-month averages of the variable of interest at lead times 2, 5, 8 and 11 months ahead of the forecast time.

The overall performance of the CCA model is documented in Fig. 11.1. Inclusion of upper ocean data in the predictor variables $\mathbf{X}(t_*)$ (Smith et al. 1995) improves the forecasts at certain times of the year but the overall CCA results in Fig. 11.1 change negligibly. The CCA predicts ENSO with a correlation between predicted and observed NINO3.4 of better than 0.6 at a lead time of 5 months.

11.5. ENSO PREDICTION USING A CONSTRUCTED ANALOGUE METHOD

Suppose we have a data library consisting of seasonal (3-month average) SST time series on a global grid for the last 45 years. Furthermore, suppose that it is June 1, 2000, and that the global SST anomaly over the last 4 seasons, namely June 1999–May 2000, is very similar to (say) the period June 1981–May 1982, i.e., a 'natural analogue' of the last year is available to us. If we assume that the global SST anomaly governs the slow evolution of ocean and atmosphere anomalies, then we could try to predict the SST anomaly for the next four seasons just by watching what happened to the SST anomaly over the four seasons following May 1982.

In practice, natural analogues are rare. Van den Dool (1994a) overcame this problem by constructing an analogue using a linear combination of all possible analogue years he had available. Specifically, van den Dool based his analysis on the above global SST anomaly time series as represented by the first five EOFs. If it is June 1, 2000, then the 'base period', would consist of the gridded global SST anomaly (using five EOFs) for March–April–May 2000 (MAM00), Dec99 – Jan00 – Feb00 (DJF00), SON99 and

JJA99. Since the global SST extends 45 years back to 1955, the number of available analogue years for the linear combination is the total number of years minus the single base year $= 1999 - 1955 = 44$ years. The constructed analogue T_{ij} for the SST anomaly at the ith grid point and the jth 3-month time interval ($j = 1, 2, 3, 4$) is the linear combination

$$T_{ij} = \sum_{k=1}^{44} \alpha_k T_{ijk} \tag{11.8}$$

where T_{ijk} is the value of the SST anomaly at the ith grid point and jth 3-month time interval for the kth analogue year. The 44 weighting coefficients α_k are determined by least squares fitting the sum in (11.8) to the corresponding data in the base period. Predictions for the global SST are then made by projecting the sum in (11.8) forward in time since we know α_k and how the values of T_{ijk} developed. Predicted NINO3.4 values can be obtained from the predicted global SST anomaly values by averaging over the region 5°S–5°N, 170°W to 120°W.

The performance of van den Dool's constructed analogue predictor is documented in Fig. 11.1. The model displays an ability to predict ENSO, the correlation of predicted and observed NINO3.4 being better than 0.6 at a lead time of 11 months.

▶ 11.6. ENSO PREDICTION USING LINEAR INVERSE MODELING

Penland and Magorian (1993) and Penland and Sardeshmukh (1995) have used 'linear inverse modeling' (LIM) both to understand and to predict Indian and Pacific Ocean SST anomalies. This technique is related to earlier work on 'Principal Oscillation Patterns' by Hasselmann (1988) and von Storch et al. (1988). I summarize the basic theory behind LIM next and then briefly discuss the model performance and ENSO physics (Section 11.6.2).

11.6.1. The basic theory

Usually when we think of modeling, we think of solving a set of equations based on physical principles. This direct or forward modeling is in contrast to 'inverse modeling' which deals exclusively with observations and uses them to extract dynamical properties from observed statistics. For example, based on inverse modeling of tropical Indian and Pacific SST anomalies, Penland and Sardeshmukh (1995) suggest that on interannual time scales, the tropical atmosphere–ocean system may be approximated by a linear system forced by spatially coherent white noise. Although the idea of inverse modeling is to deduce dynamical properties of the system, it can also be used for prediction.

To understand how LIM can be used to predict ENSO, first note that any dynamical system may be represented in the form

$$\frac{d\chi}{dt} = B\chi + \mathbf{N}(\chi) + \mathbf{F} \tag{11.9}$$

where χ is the 'state' vector having as its components all the physical variables necessary for describing the evolution of the system, B is the linear system matrix, $\mathbf{N}(\chi)$ describes the non-linear terms and $\mathbf{F}$ the forcing terms. In their model of typical SST, Penland and Magorian (1993) and Penland and Sardeshmukh (1995) assume that a subset of χ, the tropical SST, $\mathbf{T}$, is governed by a much simpler system than (11.9), namely the linear system

$$\frac{d\mathbf{T}}{dt} = B\mathbf{T} + \boldsymbol{\varepsilon} \tag{11.10}$$

where $\boldsymbol{\varepsilon}$ represents Gaussian white noise forcing.

Equation (11.10) can be solved using matrix calculus. In standard fashion, we define the exponential of a matrix by its power series expansion, e.g.,

$$\exp(Bt) = I + Bt + B^2 t^2 / 2 + \cdots \tag{11.11}$$

where I is the identity matrix. We now solve (11.10) in the same way we would for the more familiar scalar case. Specifically, multiplying (11.10) by the matrix $\exp(-Bt)$ gives

$$\exp(-Bt)\frac{d\mathbf{T}}{dt} - \exp(-Bt)B\mathbf{T} = \exp(-Bt)\boldsymbol{\varepsilon}. \tag{11.12}$$

Since B is independent of time, it follows from the power series definition of the exponential function of the matrix $-Bt$ that

$$\frac{d}{dt}(\exp(-Bt)) = -B\exp(-Bt).$$

Hence, if we write

$$A(t) = \exp(-Bt), \tag{11.13}$$

then (11.12) can be written as

$$A\frac{\mathrm{d}\mathbf{T}}{\mathrm{d}t} + \frac{\mathrm{d}A}{\mathrm{d}t}\mathbf{T} = A\boldsymbol{\varepsilon}. \tag{11.14}$$

The ith element of this matrix problem is

$$\sum_j A_{ij}\frac{\mathrm{d}T_j}{\mathrm{d}t} + \sum_j \frac{\mathrm{d}A_{ij}}{\mathrm{d}t}T_j = \sum_j A_{ij}\varepsilon_j,$$

i.e.,

$$\frac{\mathrm{d}}{\mathrm{d}t}\left(\sum_j A_{ij}T_j\right) = \sum_j A_{ij}\varepsilon_j.$$

Integration of the above equation from some general time t to $t+\Delta t$ gives

$$\sum_j A_{ij}(t+\Delta t)T_j(t+\Delta t) = \sum_j A_{ij}(t)T_j(t) + \int_t^{t+\Delta t}\sum_j A_{ij}\varepsilon_j\mathrm{d}t_*$$

which, using (11.13), can be written in the matrix form

$$\exp(-(B)(t+\Delta t))\mathbf{T}(t+\Delta t) = \exp(-Bt)\mathbf{T}(t) + \text{'noise' term.} \tag{11.15}$$

It follows from matrix multiplication of the power series expansions for $\exp(-(B)(t+\Delta t))$, $\exp(-Bt)$ and $\exp(-B\Delta t)$ that

$$\exp(-B(t+\Delta t)) = \exp(-Bt)\exp(-B\Delta t). \tag{11.16}$$

Substitution of this result into (11.15) and then multiplying first by $\exp(Bt)$ and then by $\exp(B\Delta t)$ gives

$$\mathbf{T}(t+\Delta t) = \exp(B\Delta t)\mathbf{T}(t) + \text{'noise' term.} \tag{11.17}$$

Based on (11.17) we can determine the matrix $G = \exp(B\Delta t)$ from known $\mathbf{T}$ by least squares fitting the elements of $\mathbf{T}(t)$ to $\mathbf{T}(t+\Delta t)$. Specifically, for each i, we find the coefficients G_{ij} to determine $T_i(t+\Delta t)$ from the linear combination $\sum_j G_{ij}T_j(t)$ of the $T_j(t)$. Thus, we minimize the expected value of $\left[T_i(t+\Delta t) - \sum_j G_{ij} T_j(t) \right]^2$, i.e., we minimize

$$\sum_t \left[T_i(t+\Delta t) - \sum_j G_{ij} T_j(t) \right]^2.$$

Differentiation of the above expression with respect to G_{ik} and setting the result equal to zero gives

$$\sum_t T_i(t+\Delta t) T_k(t) = \sum_j G_{ij} \sum_t T_j(t) T_k(t). \tag{11.18}$$

The above equation can be written in matrix form as

$$C(\Delta t) = GC(0) \tag{11.19}$$

where $C(\Delta t)$ is the lagged covariance matrix with elements

$$[C(\Delta t)]_{ij} = \langle T_i(t+\Delta t) T_j(t) \rangle \tag{11.20}$$

and $C(0)$ is the covariance matrix. Here the angled brackets denote a time average over the available record length and $T_k(t)$ is the value of the SST anomaly at time t at the kth grid point. The matrix B can be found from (11.19) and $G = \exp B\Delta t$. We have

$$B = \frac{1}{\Delta t} \ln \left[C(\Delta t)(C(0))^{-1} \right] \tag{11.21}$$

where the logarithm of a matrix is defined as the logarithmic power series of matrices analogous to (11.11).

Equation (11.17) suggests that if the time now is t_*, then we may predict $\mathbf{T}$ a time Δt into the future as

$$\hat{\mathbf{T}}(t_* + \Delta t) = \exp(B\Delta t)\mathbf{T}(t_*) \tag{11.22}$$

with B given as in (11.21). Penland (1989) noted that such a prediction is the most probable vector $\mathbf{T}$ at time $t_* + \Delta t$. The Linear Inverse forecast model uses the first 20 EOFs of a 3-month running mean of the monthly Indian and Pacific Ocean SST anomalies from 30°S to 30°N and from 30°E to 70°W to represent $\mathbf{T}$. This vector can be used to calculate NINO3.4 so that the LIM model can be compared with others.

11.6.2. Model performance and ENSO dynamics

Figure 11.1 shows that the LIM model's cross-verified NINO3.4 hind-cast is correlated with observed NINO3.4 at better than 0.6 for a lead time of 5 months.

As mentioned earlier, based on LIM, Penland and Sardeshmukh (1995) concluded that on interannual time scales the tropical atmosphere–ocean system may be approximated by a linear system driven by white noise. But although linear theory is very helpful in understanding ENSO, because of the dependence of anomalous deep atmospheric convection on total SST rather than just the SSTA (see Chapter 6), ENSO has at least one essential non-linearity. This non-linearity is crucial to understanding the observed tendency for ENSO events to have maximum amplitude near the end of the calendar year (see Sections 8.7 and 8.8).

11.7. COMPARISON OF STATISTICAL AND DYNAMICAL ENSO PREDICTION MODELS

The overall 'prediction' results for the dynamical models are summarized in Fig. 10.1 and for the statistical models in Fig. 11.1. In considering model performance it is worth remembering that the models are being assessed over different record lengths and different time periods. Some models are assessed on actual operational predictions, others on 'cross-verified' hindcasts and others on hindcasts that use data during the hindcast period. Bearing in mind these caveats, Figs 10.1 and 11.1 suggest that statistical and dynamical models have comparable predictive ability, the apparently better NINO3.4 forecast models being able to forecast ENSO conditions 1 year in advance with a correlation of about 0.6.

REFERENCES

Barnett, T. P., and R. Preisendorfer, 1987: Origins and levels of monthly and seasonal forecast skill for United States surface air temperatures determined by canonical correlations analysis. *Mon. Weather Rev.*, **115**, 1825–1850.

Barnston, A. G., M. H. Glantz, and Y. He, 1999: Predictive skill of statistical and dynamical climate models in SST forecasts during the 1997–98 El Niño episode and the 1998 La Niña onset. *Bull. Am. Meteorol. Soc.*, **80**, 217–243.

Barnston, A. G., and C. F. Ropelewski, 1992: Prediction of ENSO episodes using canonical correlation analysis. *J. Climate*, **5**(11), 1316–1345.

Barnston, A. G., H. M. van den Dool, S. E. Zebiak, T. P. Barnett, M. Ji, D. R. Rodenhuis, M. A. Cane, A. Leetmaa, N. E. Graham, C. F. Ropelewski, V. E. Kousky, E. A. O'Lenic, and R. E. Livezey, 1994: Long-lead seasonal forecasts – Where do we stand? *Bull. Am. Meteorol. Soc.*, **75**, 2097–2114.

Clarke, A. J., and S. Van Gorder, 2003: Improving El Niño prediction using a space – time integration of Indo-Pacific winds and equatorial Pacific upper ocean heat content. *Geophys. Res. Lett.*, **30**(7), doi: 10.1029/2002GL016673, April 10, 2003.

Hasselmann, K., 1988: PIPS and POPS – A general formalism for the reduction of dynamical systems in terms of Principal Interaction Patterns and Principal Oscillation Patterns. *J. Geophys. Res.*, **93**, 11,015–11,020.

Knaff, J. A., and C. W. Landsea, 1997: An El Niño-Southern Oscillation Climatology and Persistence (CLIPER) forecasting scheme. *Weather Forecasting*, **12**, 633–652.

Penland, C., 1989: Random forcing and forecasting using principal oscillation pattern analysis. *Mon. Weather Rev.*, **117**, 2165–2185.

Penland, C., and T. Magorian, 1993: Prediction of Niño 3 sea surface temperatures using linear inverse modeling. *J. Climate*, **6**, 1067–1076.

Penland, C., and P. D. Sardeshmukh, 1995: The optimal growth of tropical sea surface temperature anomalies. *J. Climate*, **8**, 1999–2024.

Smith, T. M., A. G. Barnston, M. Ji, and M. Chilliah, 1995: The impact of Pacific-Ocean subsurface data on operational prediction of tropical Pacific SST at the NCEP. *Weather Forecasting*, **10**, 708–714.

van den Dool, H. M., 1994a: Searching for analogues, how long must we wait? *Tellus*, **46A**, 314–324.

van den Dool, H. M., 1994b: Constructed analogue prediction of the east central tropical Pacific SST through Fall 1995. *Exp. Long-Lead Forecast Bull.*, **3**(3), 22–23.

von Storch, H., T. Bruns, I. Fischer-Bruns, and K. Hasselmann, 1988: Principal oscillation pattern analysis of the 30- to 60-day oscillation in a GCM equatorial troposphere. *J. Geophys. Res.*, **93**, 11,022–11,036.

CHAPTER TWELVE

ENSO's Influence on Marine and Bird Life

Contents

12.1. Overview

Christmas Island (1°59′N, 157°29′W) is the largest coral atoll in the world. Located in the central equatorial Pacific, it supports 18 species of breeding marine birds (Schreiber and Schreiber 1983). Schreiber and

Schreiber visited Christmas Island in November 1982, near the peak of the huge 1982 El Niño (see Figs 8.6 and 8.8). They 'discovered virtually a total reproductive failure on the island; the bird populations had essentially disappeared, and many dead and starving nestlings were present.' For example, in June 1982 there were about 10 000 pairs of Great Frigatebirds (*Fregata minor*) nested but by November 1982 there were less than 100. Most likely the fish and squid on which the birds feed had disappeared. The adults then abandoned the island and left the nestlings to starve. Such abandonment is horrible but is probably best for the survival of the species—by leaving, the adults will survive to reproduce again, perhaps many times.

The reproductive disaster at Christmas Island was one of several that occurred during the 1982 El Niño. In this chapter we will consider examples of the influence of El Niño on marine and bird life. The list is not comprehensive and many of the mechanisms are not completely understood. This is an active area of research and the number of examples of El Niño's influence has grown rapidly over the last decade.

12.2. EL NIÑO'S INFLUENCE IN THE EASTERN EQUATORIAL PACIFIC

To understand El Niño's influence in the eastern equatorial Pacific, imagine that the equatorial Pacific consists of a light surface layer of water overlying a heavier much deeper layer. In this simple model the change in density with depth is concentrated at the interface between the two layers; in practice there's a sharp change in density at the thermocline (where temperature changes rapidly with depth) and much slower changes above and below it (see, e.g., Fig. 2.7).

If no wind were blowing we would expect the sea surface and the interface between the light and heavy layers of water to be flat (see Fig. 12.1(a)). But wind does blow – the westward equatorial trade winds primarily affect the surface layer and push it westward. This thickens the surface layer in the western equatorial Pacific and thins it in the eastern equatorial Pacific (see Fig. 12.1(b)). The thinning in the east is associated with an upward displacement of the thermocline so that the thermocline is near the surface there (see Fig. 2.7). In the surface layer the zonal wind stress forcing, pushing the water westward, is balanced by the zonal pressure gradient trying to push the water back eastward to its undisturbed state (see (5.48) and (5.49)). Note that to a first approximation only the surface layer is affected by the wind and so lower layer pressures beneath the thermocline do not change when the sea level is tilted. By the physical argument given at the

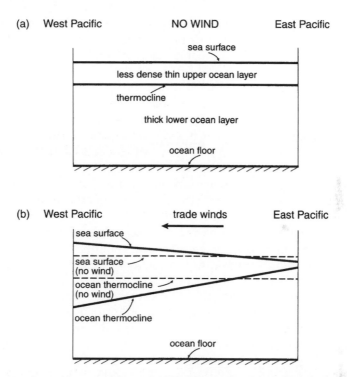

Fig. 12.1 (a) A zonal equatorial section showing a two density layer approximation to the equatorial ocean when there is no wind. The layer depths are not drawn to scale – in practice the lower layer depth is about 40 times the upper layer depth. (b) The westward trade winds push the very thin upper layer westward, thickening it in the western Pacific and thinning it in the eastern Pacific. The upward sea level and downward thermocline displacements are not to scale – in practice the sea level is only about 40 cm higher in the west but the thermocline is about 100 m deeper (see text).

end of Section 3.3.2, the thermocline displacement shown in Fig. 12.1 is actually very much greater than the sea level displacement.

The thinning of the surface layer in the east is of considerable importance to marine life. Phytoplankton (drifting, mostly single-celled marine plants) are at the base of the food chain. As plants they need light and nutrients. Since light does not penetrate far into the ocean, for plant-life to exist the nutrients must be near the surface. The surface layer is nutrient-poor but in the eastern equatorial Pacific the nutrient-rich lower layer is shallow enough on average to be lit so photosynthesis by the phytoplankton is possible. This leads to an abundant plant and animal life.

However, an El Niño can devastate this balance. As noted in Chapter 2, during an El Niño the equatorial wind does not blow as strongly westward.

Consequently, the interface between the two ocean layers is not as tilted, the nutrients are not as close to the surface in the eastern equatorial Pacific, the phytoplankton are not as plentiful and the higher levels of the food chain are affected. This was probably the reason for the reproductive failure of sea birds on Christmas Island during the Fall of 1982. Then there were strong westerly wind anomalies and, as the thermocline became less tilted anomalously warm nutrient-poor surface water surrounded Christmas Island. This would lead to fewer phytoplankton and, further up the food chain, to fewer small fish and squid and consequently to a devastation of the bird life on the island.

This severe El Niño also devastated the fur seal population on the Galapagos Islands more than 7000 km along the equator to the east. There all Galapagos fur seal pups in the study area that were born in 1982 had died of starvation before they were 5 months old (Limberger et al. 1983). They starved because their foraging mothers apparently could not find enough food to feed themselves and so produce enough milk for the pups. Normally the mothers' foraging trips to sea last about 1.5 days on average but in November and December 1982 the average trip was 5.1 days. It is likely that the deeper thermocline and influx of warm, nutrient-poor surface water resulted in fewer phytoplankton and consequently fewer fish and squid prey for the mother seals.

12.3. Peruvian Anchovies and Guano Birds

In 1967 approximately one-fifth of the world's fish catch came from a narrow coastal strip of water 800 miles long and 30 miles wide, i.e., about 0.02% of the surface area of the world ocean. In this Peruvian coastal region, upwelling, or the vertical upward movement of water, takes place. The source of the upwelled water is about 40–80 m below the surface (Brink et al. 1983). Because the thermocline is usually near the surface in the eastern equatorial Pacific (see Fig. 12.1(b) and Fig. 2.7), the source of the upwelled water is deep enough to be rich in plant nutrients. As explained previously, nutrients in the well-lit surface layer support photosynthesis by phytoplankton and the huge fish population.

Why does the coastal upwelling occur? Winds typically blow parallel to the Peruvian coast and exert an along-coast northwestward stress on the surface water. This force must be balanced by an opposite southeastward force or the water will continually accelerate. On a rotating earth this southeastward force is the Coriolis force. The Coriolis force acts at 90° to the left of a current in the Southern Hemisphere, so for there to be a southeastward Coriolis force, there must be a transport of surface water

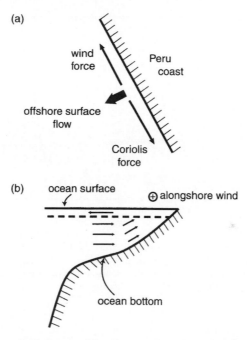

Fig. 12.2 (a) The main balance of forces near the ocean surface along the Peruvian coast as seen from above. The effect of the alongshore wind is to drive an offshore transport of surface water so that the wind force can be balanced by the Coriolis force. (b) A section perpendicular to the Peruvian coast showing the alongshore wind (into the paper), the offshore wind driven surface flow (from the dashed line to the ocean surface) and the compensating flow beneath the surface to conserve the mass lost offshore. Near the coast the compensating flow must have an upwelling component because of the tilted ocean floor.

offshore and perpendicular to the coast (see Fig. 12.2(a)). In other words, the northwestward wind stress drives an offshore near-surface transport of water (called the 'Ekman transport') because then the northwestward wind force can be balanced by the Coriolis force. To conserve mass, this offshore surface transport, in a layer about 10–30 m deep, is approximately balanced by onshore subsurface flow in a layer at about 40–80 m depth (Brink et al. 1980; Smith 1981). The shoaling ocean floor causes this onshore flow to upwell (see Fig. 12.2(b)).

As mentioned earlier, the Peruvian coastal upwelling results in abundant marine life. The main commercial fish caught off Peru is the Peruvian anchovy (*Engraulis ringens*). Most of the catch is converted to fishmeal and sold to developed nations as a food supplement for chickens and pigs. The fishery officially began in the early 1950s and, aided by the collapse

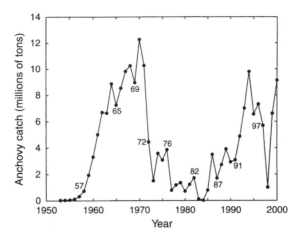

Fig. 12.3 Peruvian annual anchovy catch with El Niño years labeled. The catch usually falls during the El Niño year or the next year (see, e.g., 82–83 and 97–98) since El Niño's effects often extend into the following year (the Peruvian Marine Institute IMARPE generously supplied the data for this figure).

of the California sardine fishery in the 1950s, it developed rapidly (see Fig. 12.3). The California sardine fishery collapse helped the Peruvian fishery in two main ways: first, the idle and highly efficient California fishmeal processing plants and fishing vessels were available for sale to Peru; second, the Peruvian anchovy fishmeal found a ready market in the United States since the California sardine fishmeal was not available. So rapid was the development of the Peruvian anchovy fishery that by the mid-1960s Peru had become the number one fishing nation on earth.

But by this time the number of boats and processing plants was bigger than the fishery could sustain, especially when El Niño is taken into account. As explained earlier, when an El Niño occurs the westward equatorial winds on average blow less strongly westward and the thermocline and associated nutricline in the eastern equatorial Pacific deepens. Coastal upwelling still occurs along the Peruvian coast but, since the thermocline and nutrients are deeper than 80 m, the upwelled water is poor in nutrients, the phytoplankton biomass decreases and stresses the anchovy population which feeds on it. In 1972, with the Peruvian fishery already stressed by over-fishing, the strong 1972 El Niño hit and the anchovy catch fell by more than 50% (see Fig.12.3). For the next two decades the anchovy catch remained low, being essentially zero in 1983 and 1984 when fishing was almost completely closed following the huge 1982–1983 El Niño. Except for the 1998 catch which was low due to the huge 1997–1998 El Niño, Fig. 12.3 shows that in the 1990s the Peruvian anchovy catch has recovered to be nearly as large as it was in the late 1960s/early 1970s. Improved management regulations after the huge 1982 El Niño have led to this recovery (Jahncke 1998).

Although the anchovies have recovered, the same cannot be said of the guano birds which depend mostly on anchovies for food. These birds, named after their droppings, or guano, include cormorants, boobies and pelicans. Figure 12.4 shows that their current population is a small fraction of what it was in the 1950s. The population decreased during the 1957, 1965, 1969, 1972, 1982–1983 and 1997 El Niños because of the decrease in the anchovy population then. The bird population probably will not recover to the level of the early 1950s because the Peruvian fishery is harvesting the fish the birds once ate. The fishery is likely to continue since it is economically more valuable than the guano fertilizer produced by the birds.

The response of the guano birds and the anchovy catch to the 1957 and 1965 El Niños suggest that the fishermen gained control of the anchovy catch between these years. The 1957 El Niño was large and the guano birds population fell sharply, presumably due to the absence of the anchovies. The fishermen's 1957 anchovy catch was negligibly affected because it was so small and because the fishermen were better anchovy predators than the birds. In 1965, a moderate El Niño, the bird population fell sharply again because by then the fishermen were taking much of the catch and what was left over was drastically reduced under El Niño conditions. The sudden

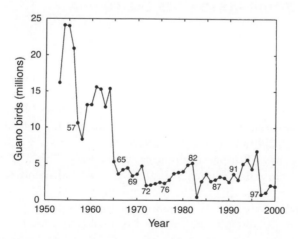

Fig. 12.4 Estimated annual Peruvian guano bird population with El Niño years labeled. (The Peruvian Marine Institute IMARPE generously supplied the data for this figure.)

drop in bird population in 1965 was a warning that the fishermen's catch should be managed. Biologists alerted the government to the problem but their advice was ignored.

Although El Niño is bad news for Peruvian anchovies, guano birds and other Peruvian marine life, some marine populations thrive. For example, during the severe El Niño conditions in 1983, while many commercially important invertebrates such as mussels, clams, crabs, limpets, false abalone and sea urchins died or disappeared, the 1983 Peruvian scallop (*Argopecten purpuratus*) catch increased about 40-fold (Wolff 1984). The extraordinary growth and reproduction by the scallops was probably due to two main factors (Barber and Chavez 1986; Wolff and Mendo 2000). Firstly, the elevated temperatures led to faster maturation of the juvenile and adult scallops and faster development (and therefore less dispersion from the scallop banks) at the larval stage. Secondly, because of the El Niño, there were fewer predators and fewer competing filter feeders.

In addition to the increased scallop catch, during the height of the 1982–1983 El Niño, fishermen reported increased landings of tropical oceanic fish species as the thermocline deepened and warm oceanic tropical waters invaded the Peruvian coastal region (Velez et al. 1984). The Peruvian coastal city of Callao (12°S), recorded 51 species of fish, 6 species of shrimp and 4 species of crabs normally found to the north or in oceanic waters (Velez et al. 1984).

12.4. ZOOPLANKTON OFF CALIFORNIA

12.4.1. Background

At the peak of the California sardine fishery in 1936, fishermen harvested 700 000 tons but from 1945 to 1947 the annual catch fell from 550 000 to 100 000 tons. Shocked by the disappearance of the sardines, several institutions began collaborative research in 1949. Named the California Cooperative Fisheries Investigations (CalCOFI) in 1953, the goal of the collaboration was to determine the cause of the collapse – was it due to overfishing, biological interactions within the region or external environmental influences? Over the last half-century CalCOFI's scope has broadened to include the influence of the physical, chemical and biological environments on many species besides sardines. The five-decade long multi-disciplinary data set is probably unique for its longevity and detail.

The sardine population fluctuates on a time scale much longer than El Niño and we will not discuss this further, except to say that overfishing was probably not the sole cause for the collapse – by counting sardine scales in varved anaerobic sediments, Soutar and Isaacs (1969, 1974) showed that extreme fluctuations in the sardine population occurred long before the twentieth century fishery existed. Many California current populations like rockfish (see the summary by Lenarz et al. 1995) and birds (e.g., Brandt's Cormorants, see Nur and Sydeman 1999) are affected by El Niño and the CalCOFI data set has enabled researchers to examine them. We will focus on California zooplankton near the base of the food chain. But first we must understand how El Niño influences California coastal waters.

12.4.2. El Niño and California coastal waters

Why should El Niño influence California coastal marine populations many thousands of miles from the equator? Consider again the idealized two-layer ocean in Fig. 12.1. At the low frequencies corresponding to El Niño, the flow is nearly geostrophic. Under the geostrophic approximation, ocean pressure must be constant along the coast because, if it were not, a pressure difference would cause a flow into the coast which is impossible since the coast is essentially impermeable. But if the pressure on a given level surface in the ocean is constant, the sea surface must be level, i.e., the sea level height (and pycnocline depth) in the two-layer model ocean must be constant along the coast. Thus, during El Niño, the raised sea level and depressed thermocline at the equator is communicated all along the western coast of the Americas, including southern California. As in the Peruvian ecosystem discussed earlier (see Section 12.1), the depressed thermocline implies that the nutrients are not as close to the well-lit surface layer. Thus even though the southern California winds continue to favor upwelling, the waters upwelled are lower in nutrients, there are fewer phytoplankton and marine life suffers.

Evidence that the El Niño signal is transmitted oceanographically along the western coast of the Americas has been provided for sea level by Enfield and Allen (1980), Chelton and Davis (1982) and Pizarro et al. (2001), for sea surface temperature by Enfield and Allen (1980) and for 20°C isotherm displacement by Kessler (1990). The El Niño signal is not transmitted instantaneously along the coast but instead propagates poleward. A detailed theoretical discussion of low-frequency flow dynamics along eastern ocean boundaries is given in Section 4.3. While interannual variations in coastal wind stress do not appear to influence the upwelling and marine life along much of the west coast of the Americas, Enfield and Allen (1980) showed that interannual coastal wind stress does influence sea level and thermocline displacement north of southern California.

12.4.3. Rossby waves and ENSO currents off the California coast

Section 4.3 discussed the dynamics of low-frequency flow near the eastern boundary of the Pacific Ocean. On ENSO time scales, linear theory predicts (see Sections 4.3.3, 4.3.4 and 4.3.7) that long, non-dispersive Rossby waves propagate westward from the coast. Recently, using the long record of CalCOFI hydrographic observations, Clarke and Dottori (2007) showed that monthly sea level anomalies (departures from the seasonal cycle) propagate westward from the coast of California, qualitatively consistent with theory. The westward propagation speed, confirmed by an analysis of independent satellite altimeter data to be approximately $4.2\,\mathrm{cm\,s^{-1}}$, was about double that expected theoretically. Based on this result, Clarke and Dottori (2007) showed that it is possible to relate coastal sea level anomalies to low-frequency anomalies in the California Current and to large fluctuations in the zooplankton population. We discuss this theory below.

Since the sea level anomaly η' propagates westward from the coast, we may write that

$$\eta' = G(x + \gamma t) \tag{12.1}$$

for some general function G. In (12.1) $\gamma = 4.2\,\mathrm{cm\,s^{-1}}$, x is distance eastward and t is the time. Differentiating (12.1) with respect to x and then separately with respect to time shows that

$$\frac{\partial \eta'}{\partial t} = \gamma \frac{\partial \eta'}{\partial x}. \tag{12.2}$$

We can use (12.2) to calculate the large-scale low-frequency surface currents in the California Current. At the low frequencies of interest the currents are nearly geostrophic, so that if n is the perpendicular distance into the coast and s is alongshore (see Fig. 4.8), the surface currents u' perpendicular to the coast and v' alongshore are

$$u' = \frac{-\partial p'}{\partial s}\bigg/(\rho_* f) = -g\frac{\partial \eta'}{\partial s}\bigg/f \tag{12.3}$$

and

$$v' = \frac{\partial p'}{\partial n}\bigg/(\rho_* f) = g\frac{\partial \eta'}{\partial n}\bigg/f. \tag{12.4}$$

The second equality in (12.3) and (12.4) follows because the pressure anomaly at $z = 0$ is $\rho_* g \eta'$. The anomalous large-scale flow is almost completely alongshore so $\partial \eta'/\partial s \ll \partial \eta'/\partial n$. Therefore if $\mathbf{i}$ is the unit eastward vector,

$$\frac{\partial \eta'}{\partial x} = \mathbf{i} \cdot \nabla \eta' = \mathbf{i} \cdot \left(\mathbf{e}_n \frac{\partial \eta'}{\partial n} + \mathbf{e}_s \frac{\partial \eta'}{\partial s} \right) \approx \mathbf{i} \cdot \mathbf{e}_n \frac{\partial \eta'}{\partial n} \tag{12.5}$$

where, as in Chapter 4, $\mathbf{e}_n$ and $\mathbf{e}_s$ are unit vectors in the direction of increasing n and s, respectively. Since θ is the angle between $\mathbf{e}_n$ and $\mathbf{i}$ (see Fig. 4.8), (12.5) simplifies to

$$\frac{\partial \eta'}{\partial x} = \cos \theta \frac{\partial \eta'}{\partial n} \tag{12.6}$$

and so, from (12.4)

$$v' = g \frac{\partial \eta'}{\partial x} \Big/ (f \cos \theta) \tag{12.7}$$

or, from (12.2)

$$v' = g \frac{\partial \eta'}{\partial t} \Big/ (f \gamma \cos \theta). \tag{12.8}$$

Notice that, using (12.8), we can estimate the anomalous large-scale low-frequency interannual surface flow anywhere in the California Current using the time derivative of local interannual sea level. Since local sea level is related to coastal San Diego sea level by westward wave propagation, large-scale interannual variations in the California Current can be estimated by appropriately lagging the time derivative of interannual San Diego sea level.

Closely associated with the alongshore current anomaly is the anomalous alongshore particle displacement $(\Delta s)'$ (see also (4.59) and (4.60)). Since, by (12.8)

$$\frac{d}{dt}(\Delta s)' = v' = g \frac{\partial \eta'}{\partial t} \Big/ (f \gamma \cos \theta), \tag{12.9}$$

an integration with respect to time shows that

$$(\Delta s)' = g \eta' \Big/ (f \gamma \cos \theta). \tag{12.10}$$

From Fig.4.8 the alongshore particle displacement is poleward when $(\Delta s)' > 0$ and equatorward when $(\Delta s)' < 0$. By the eastern ocean boundary dynamics discussed earlier in Section 12.4.2, the anomalously high eastern equatorial Pacific sea level during an El Niño will usually result in $\eta' > 0$ at the California coast. A typical η' peak during El Niño $\approx +5$ cm, so, with $g = 9.8 \, \text{ms}^{-2}$, $\gamma = 4.2 \, \text{cm s}^{-1}$, $f = 8 \times 10^{-5} \text{s}^{-1}$ and $\theta = 46°$, from (12.10) the alongshore particle displacement ≈ 200 km. Taking into account the equatorward displacement due to $\eta' < 0$, anomalous alongshore particle displacement has a range of about 400 km. The corresponding anomalous velocity v' can be estimated from (12.9) for a typical El Niño frequency ω as $\omega (\Delta s)'$. For $(\Delta s)' = 200$ km and $\omega \sim 2\pi/3$ years, v' is only about 1–$2 \, \text{cm s}^{-1}$. Physically, even though the velocity is small, the alongshore displacement is large because the weak current is in one direction for a long time.

12.4.4. Zooplankton population and El Niño

In a seminal study Chelton et al. (1982) used CalCOFI data to show that interannual zooplankton fluctuations in the California Current are not related to variations in the local alongshore wind forcing and coastal upwelling, but rather to variations in the California Current. They noted that in the mean the nutrient content of the surface water is higher in the north than the south, and assumed that when the sea level is anomalously low, the equatorward California Current is stronger than normal and transports the higher nutrient northern water southward. The higher nutrient water then results in more phytoplankton and zooplankton.

One difficulty with the above mechanism is that it does not take into account Rossby wave dynamics. For example, when the sea level is anomalously low and at a minimum, $\partial \eta'/\partial t$ is zero. Based on geostrophy and Rossby wave dynamics, (12.8) then shows that $v' = 0$, i.e., there is no anomalous flow. Therefore the California Current is not in fact stronger southward when the sea level is anomalously low. However, (12.10) tells us that η' is proportional to alongshore particle *displacement* so that when sea level is anomalously low a higher nutrient water particle north of the California Current region should be displaced about 200 km southward into it. This makes higher nutrient water available locally, leading to more than usual phytoplankton and zooplankton there. Conversely, when $\eta' > 0$, lower nutrient water particles are displaced northward into the California Current region, the nutrient content of the water is lower and the phytoplankton and zooplankton populations are likely to fall.

How, mathematically, should we expect η' and the zooplankton concentration to be related? If the sea level rises by an amount $\delta \eta$, then by

(12.10) we should expect an anomalous displacement of lower nutrient water from the south into the California Current region. In addition, as noted earlier near the end of Section 3.3, increased sea level corresponds to a deeper thermocline and fewer nutrients. Thus a rise $\delta\eta$ in sea level should lead to a decrease $-\delta P$ in the zooplankton population P. The decrease $-\delta P$ should also be proportional to P because, for a given decrease in nutrient concentration, the drop in zooplankton population $-\delta P$ will be larger if there are more zooplankton that suffer. For this simple model, in which $-\delta P$ is proportional to both $\delta\eta$ and P, we write

$$\delta P = -\lambda \delta \eta P \qquad (12.11)$$

where $\lambda > 0$ is a constant of proportionality. Dividing both sides of (12.11) by P and integrating from a sea level η_0 when $P = P_0$ to general η and P gives

$$\ln P - \ln P_0 = -\lambda(\eta - \eta_0). \qquad (12.12)$$

Subtracting the mean and annual cycle of both sides of (12.12) from (12.12) leads to the anomaly equation

$$(\ln P)' = -\lambda \eta'. \qquad (12.13)$$

The relationship (12.13) applies locally but cannot be tested point by point because the zooplankton data are too noisy locally. However, we might expect that if $(\ln P)'$ is averaged over the entire CalCOFI region, then $(\ln P)'$ should be proportional to $-\eta'$ averaged over the CalCOFI region. Clarke and Dottori found that these two time series were indeed correlated with correlation coefficient $r = 0.67$.

The left-hand side of (12.12) is $\ln(P/P_0)$ so if we take exponentials of both sides of (12.12) we have

$$P = P_0 \exp(-\lambda(\eta - \eta_0)), \qquad (12.14)$$

i.e., P is very sensitive to changes in sea level η. This hypersensitivity of P to fluctuations in η on El Niño time scales also occurs on longer time scales. For example, in the early 1990s the estimated California zooplankton population was only about one-quarter of what it was in the early 1970s (Roemmich and McGowan 1995a,b). Clarke and Lebedev (1999) have proposed that, like the El Niño case, these decadal fluctuations originate from variations in the equatorial east–west winds. For example, during the 1970s and 1980s the equatorial easterlies weakened and the thermocline in the eastern equatorial Pacific consequently deepened. Because of geostrophic balance the thermocline also deepened all along the Americas to California and the California zooplankton population, hypersensitive to the changed thermocline depth, fell dramatically.

12.5. THE LEAKY WESTERN EQUATORIAL PACIFIC BOUNDARY AND AUSTRALIAN 'SALMON'

12.5.1. Physical background

As explained in Section 12.2, during an El Niño equatorial westerlies reduce the overall easterly trade wind forcing and consequently the normal upward sea level slope toward the west along the equator is reduced. The sea level is then anomalously low in the western equatorial Pacific and anomalously high in the eastern equatorial Pacific. Because of gaps in the western equatorial Pacific boundary, sea level is anomalously low around western New Guinea. Consistent with geostrophic balance and no normal flow into the coast at interannual frequencies, sea level should be spatially constant and lower than normal all along the west coasts of New Guinea and western and southern Australia (see Fig. 12.5). Similarly, during La Niña the equatorial wind anomalies reverse to become anomalously easterly, making sea level anomalously high in the western equatorial Pacific and around Australia's western and southern coasts. Pariwono et al. (1986) first discovered that there was an ENSO sea level signal around Australia's western and southern coasts and Clarke (1991) suggested the above dynamics to explain it (see also the earlier discussion in Section 4.2.3).

Australia's western coastline has a large north–south component but part of its southern coastline is nearly east–west for over 1000 km. As discussed earlier in Section 4.3.7, coastline direction fundamentally affects the interannual coastal flow. Specifically, along Australia's western coastline, the theory of Chapter 4 suggests that the coastal ENSO signal should propagate westward into the interior as long Rossby waves. However, along part of Australia's southern coastline particles of water do not experience a change in Coriolis parameter because the flow is parallel to the coast and therefore east–west. Consequently, the Rossby wave mechanism fails and f-plane dynamics applies. When f-plane dynamics applies, the flow in this remotely forced ocean signal should behave like a coastal Kelvin wave with sea level proportional to alongshore velocity (see Sections 4.3.2–4.3.7).

Li and Clarke (2004) found that satellite estimates of sea level height are consistent with the above theory. Specifically, the nearly zonal southern continental shelf edge interannual sea level falls rapidly with distance from

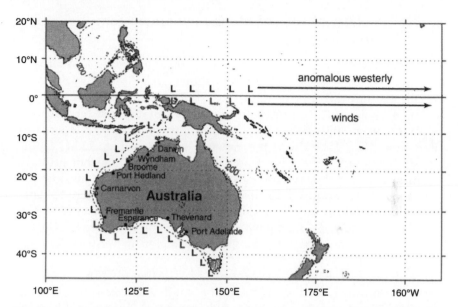

Fig. 12.5 Western equatorial Pacific showing the 200 m isobath (dashed line) around Australia and the anomalous low sea level (*L*) resulting from the anomalous equatorial westerly winds which blow in the central equatorial Pacific between about 160°E and 150°W during El Niño. Anomalous equatorial easterly winds blow during La Niña and the sea level around western and southern Australia is usually higher than normal. (Adapted from Li and Clarke 2004.)

the shelf edge with the coastal Kelvin wave radius of deformation scale while off Australia's northwest coast interannual sea level propagates westward from the coast and the sea level gradient is very weak. On ENSO time scales currents are geostrophic, so the speed of the flow is proportional to the sea level gradient. Therefore the nearly zonal southern shelf edge flow is quite strong, varying typically from about $10 \, \mathrm{cm\,s^{-1}}$ westward to $10 \, \mathrm{cm\,s^{-1}}$ eastward. During El Niño the sea level is usually anomalously low at the coast (see Fig. 12.5) so the along-shelf edge anomalous flow has anomalously higher sea level south of the shelf edge, i.e., the anomalous geostrophic flow is westward. During La Niña, the equatorial winds are anomalously easterly instead of westerly, the sea level is anomalously high in the western Pacific and around western and southern Australia, and the anomalous geostrophic flow along the southern coast is eastward.

The anomalous flow has a strong influence on the population of western Australian salmon (Lenanton et al. 1991; Li and Clarke 2004) and we will consider this next. The discussion follows closely that given in Li and Clarke (2004).

12.5.2. The effect of coastal ENSO flow on western Australian salmon

Although western Australian salmon (*Arripis truttaceus*) looks like a salmon, it is really a perch. During March and April it spawns off Australia's southwestern coast (Fig. 12.6). Some larvae settle near south-western Australia but many are carried eastward by the eastward-flowing Leeuwin Current and settle about 3–6 months later in protected coastal waters more than 1000 km to the east (see arrows in Fig. 12.6). At the start of their journey the salmon are only millimeters long but grow to 5–8 cm by the time they reach the eastern nursery areas. The juvenile fish spend about 6–12 months in the inshore nurseries and then swim to more exposed coasts where they aggregate in large schools seaward of the surf zone. After several years, during which time the salmon can grow to 80 cm long, the salmon head back to the spawning region in western Australia and begin the life cycle again.

Based on the discussion in Section 12.5.1, the ENSO flow along Australia's southern coast strengthens the south coast eastward-flowing Leeuwin Current during a La Niña and weakens it during an El Niño. Since the salmon eggs and larvae are being advected by the flow from March to September, recruitment should depend on the strength of the current during this time. As noted in Section 12.5.1, the ENSO flow is proportional to coastal sea level and, because the interannual sea level at Port Adelaide is representative of sea level all along the Australian coast (see Fig. 12.5), the anomalous transport from March to September should

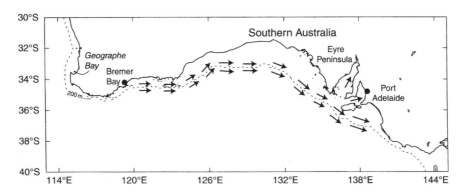

Fig. 12.6 The southern Australian coast and 200 m isobath. Western Australian salmon spawning mainly occurs between Geographe Bay and Bremer Bay during March and April. Some larvae settle locally, but many eggs and larvae are advected eastward to the protected coastal nursery grounds east of Eyre Peninsula. (Adapted from Li and Clarke 2004.)

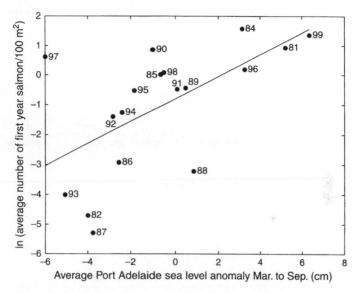

Fig. 12.7 Relationship between the pre-recruit indices (ln (average number of first year salmon)/100 m²) for Australian salmon in SA waters, 1981–1999 with 1983 missing, and the average of the sea level anomaly from March to September at Port Adelaide in that year. (Adapted from Li and Clarke 2004.)

be well represented by anomalous Port Adelaide sea level from March to September. Figure 12.7 shows that there is an approximate positively correlated ($r = 0.6$) linear relationship between this transport index and an index of salmon recruitment. The positive correlation is what we would expect physically: higher coastal sea level corresponds to a stronger eastward flow and more recruits, while lower sea level corresponds to weaker eastward flow and fewer recruits.

The recruitment index is logarithmic, and so, as for the zooplankton off California (see Section 12.4.4), salmon recruitment is exponentially dependent on anomalous sea level and therefore extremely sensitive to it. Over the observed sea level anomaly range recruitment can vary by a factor of 100. The exponential dependence can be explained by a similar argument to that used for the California zooplankton.

Specifically, the increase δn in the number of salmon recruited in a given year will depend on how much stronger the current is that year from March to September. We expect δn to be proportional to the increase in current and therefore to be proportional to $\delta \eta$, the increase in coastal sea level. We also expect δn to be proportional to the number of salmon being transported since if the number transported (say) doubles then so, roughly, should the increased recruitment. But the number transported should be highly correlated with the number n recruited so δn should be approximately proportional to n.

Based on the above, we expect δn to be proportional to both $\delta \eta$ and n and so

$$\delta n = \mu n \, \delta \eta \qquad (12.15)$$

for some constant μ. Dividing both sides by n and integrating from some neutral state when $n = n_0$ and $\eta' = 0$ to some general state gives

$$\int_{n_0}^{n} \frac{dn}{n} = \mu \int_{0}^{\eta'} d\eta,$$

i.e.,

$$\ln \left(n / n_0 \right) = \mu \eta'. \qquad (12.16)$$

In practice the number n recruited is measured by the average number of first year salmon from a $100 \, \text{m}^2$ water sample. So based on (12.16) we expect $\ln n$ to be linearly related to η'. Figure 12.7 shows that this is approximately true.

12.6. ROCK LOBSTERS AND THE LEEUWIN CURRENT OFF WESTERN AUSTRALIA

12.6.1. Physical background

One of the world's strangest currents is the Leeuwin Current. Although the wind along Australia's western coast from its southwestern tip to 22°S blows northward along the coast, the Leeuwin Current flows strongly southward along the edge of the shelf in the opposite direction to the wind! Why does it do this?

As explained in Section 12.2, because of the mean Pacific equatorial trade winds, in the mean the thermocline in the western equatorial Pacific is deep and the sea level high. Unlike the western equatorial Atlantic and Indian Oceans, the western equatorial Pacific boundary is broken and irregular (see Fig. 12.8), resulting in a major leak into the tropical Indian Ocean and a mean high sea level and a deep thermocline there. This high sea level extends westward south of the latitude of southern Java ($\approx$ 9°S). This leaky western boundary physics is the same as that discussed in Section 12.5 for El Niño/La Niña except that now we consider the

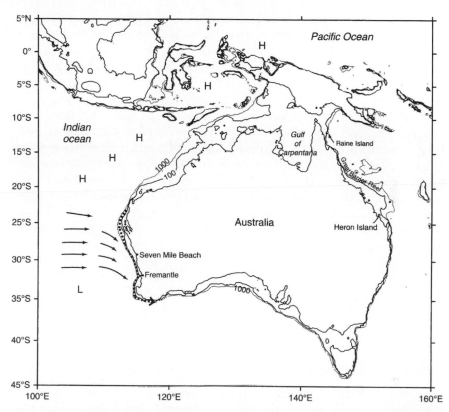

Fig. 12.8 Australia and vicinity showing key features related to the Gulf of Carpentaria banana prawns, the Western Australian rock lobsters and the Great Barrier Reef green turtles. The 'H' in the western equatorial Pacific and off the west coast of Australia refers to the relatively high mean sea level communicated from the western equatorial Pacific through the gappy 'boundary' between the Pacific and Indian Oceans. The 'L' refers to the relatively low mean sea level in the southern Indian Ocean. The arrows toward the western Australian coast show the geostrophic onshore flow feeding the Leeuwin current which can be seen running along the western Australian shelf edge and continuing along the shelf edge of southern Australia. Fremantle and Seven Mile Beach are, respectively, sea level and puerulus measurement sites (see Fig. 12.8) while Heron Island (southern Great Barrier Reef) and Raine Island (northern Great Barrier Reef) are measurement sites for nesting green turtles.

response to the mean trade winds rather than the fluctuating equatorial ENSO zonal wind anomalies.

Because of the leaky western Pacific, the sea level off the northwest Australian coast is higher than that in the south (see, e.g., Godfrey and Ridgway 1985) and so the ocean pressure is higher in the north than the south. But if the mean ocean pressure is higher in the north then, by geostrophic balance, there must be a mean eastward flow toward Australia (see Fig. 12.8). When this flow reaches Australia's continental shelf edge, it turns southward as the swiftly flowing Leeuwin Current. Since this current is in geostrophic balance, the sea level is higher near the coast and, in terms of the two-layer model (see Section 3.3.2), the thermocline deepens near the coast (see (3.45)).

The deepened thermocline has profound implications for Australia's west coast fisheries. As we have already seen in Section 12.3 the equator-ward wind stress along the Peruvian coast results in a surface transport of water offshore, the upwelling of nutrient rich waters to the surface and abundant marine life. When an El Niño occurs, the thermocline is dis-placed downward and, although the wind continues to blow, the upwelled waters are above the thermocline, are poor in nutrients and marine life suffers severely. Off western Australia, particularly in the southern summer, the along-coast winds are favorable for coastal upwelling but, because of the deepened coastal thermocline the upwelled waters are nutrient-poor and the luxuriant coastal marine life seen in Peru is absent. Essentially, coastal western Australia has a permanent El Niño! And yet, in spite of this, Australia's No. 1 single species fishery is that of the western rock lobster whose life cycle depends on Leeuwin Current dynamics.

12.6.2. The life cycle of the western rock lobster

Figure 12.9 helps summarize the life history of the western rock lob-ster. The lobsters spawn on the outer continental shelf between July and September and the larvae hatch in the Southern Hemisphere summer between December to March. They rise to the surface, are carried rapidly offshore in the thin wind-driven surface Ekman layer and widely dis-persed over the southeastern Indian Ocean. The greatest concentrations are between about 400 and 1000 km from the coast.

By the southern winter and spring the larvae have reached mid and late larval stages, remaining at depths of 50–120 m during the day. At this depth they are below the weak winter and spring wind-driven surface Ekman layer flow and right in the heart of the eastward geostrophic flow that is responsible for the Leeuwin Current. The eastward flow carries the larvae toward the current where they metamorphose to the pueru-lus stage. The pueruli can swim and head toward the coast across the continental shelf in spring and summer. There they settle in the coastal shallow reefs. After about 4 years of growth in these reefs, most juvenile lobsters migrate offshore into depths of 30–200 m. In about another 1–2 years the juveniles have become adults and can spawn and begin the life cycle again.

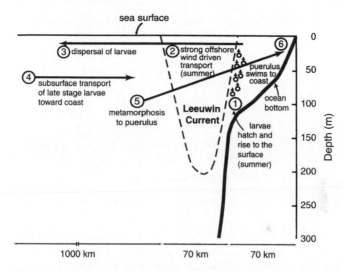

Fig. 12.9 Summary of the oceanic cycle of the western rock lobster. The lobsters spawn on the outer continental shelf and (1) the larvae hatch in the southern summer and rise to the surface. The strong wind-driven near-surface Ekman transport carries the larvae rapidly offshore (2) and they are widely dispersed (3) over the southeastern Indian Ocean. In the winter and spring the onshore subsurface geostrophic transport carries the larvae toward the coast (4) where they metamorphose to the puerulus stage (5). The pueruli can swim and they head toward the coast where they settle in the coastal reefs (6). The dashed line shows the approximate position of the Leeuwin Current. (Based on Pearce and Phillips (1988) and Smith et al. (1991).)

12.6.3. Prediction of the rock lobster catch

Phillips (1986) worked out a way to predict the rock lobster catch 4 years in advance. Fishermen catch both mature and immature lobsters, much of the catch being the juveniles emigrating from the coast. The number of juveniles emigrating depends directly on the pueruli which settle in the shallow reefs typically 4 years earlier. Using lobster collectors of artificial seaweed, Phillips was able to estimate a 'settlement index' of the pueruli and hence estimate the lobster catch 4 years later.

12.6.4. Variations of the rock lobster catch and ENSO

The puerulus annual settlement index, key to estimating the rock lobster catch, often halves or doubles from one year to the next. What causes this variability? Pearce and Phillips (1988) showed that the natural logarithm of the annual settlement index is highly correlated (correlation coefficient $r = 0.73$) with the annual sea level at Freemantle, a port on Australia's

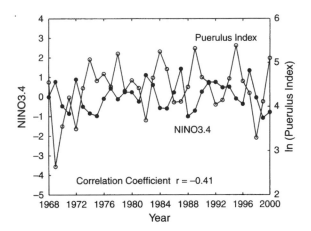

Fig. 12.10 The natural logarithm of the annual settlement index of rock lobster pueruli (open circles) and an annual average (Jan–Dec) of the El Niño index NINO3.4 (solid circles). The settlement index is defined as the average number of pueruli per artificial seaweed collector from May of one year to April of the next for collectors at Seven Mile Beach (see Fig. 12.8). The settlement season 1968/1969 is plotted as 1968, 1969/1970 as 1969, etc. Dr. Chris Chubb of the Department of Fisheries in Western Australia generously provided the settlement data. The NINO3.4 data were obtained from the website of the Climate Prediction Center of the National Center for Environmental Prediction, USA (http://www.cpc.ncep.noaa.gov/data/indices/).

southwestern coast (see Fig. 12.8). Actually, as explained in Section 12.5.1, interannual sea level, highly correlated with El Niño, is observed all along Australia's western and southern coastlines (see Fig. 12.5). This suggests that the settlement index and ENSO are highly correlated and they are (see Fig. 12.10). Why, physically, should this be so?

An analysis by Clarke & Li (2004) may provide an explanation. Their theory suggests that the strong ($\approx 0.4\,\mathrm{m\,s^{-1}}$) and narrow ($\approx 100\,\mathrm{km}$ wide) Leeuwin Current discussed earlier in this section is strong enough to invalidate the assumption that the interannual flow can be regarded as a perturbation to a background state of rest. Consequently the low-frequency long Rossby wave theory discussed earlier in Chapter 4 (see (4.41)) is invalid and instead a narrow interannual flow, with the same width as the Leeuwin Current and varying in time like the ENSO coastal sea level, is predicted. Hydrographic data at 32°S (Feng et al. 2003) and satellite altimeter observations (Clarke and Li 2004) confirm the existence of this narrow ENSO flow which hugs the shelf edge in the same location as the Leeuwin Current. When the sea level is higher than normal at the coast, typically during a La Niña, the ENSO flow anomaly is southward and the Leeuwin Current strengthens; when the coastal sea level is lower than

normal at the coast, typically during an El Niño, the ENSO flow anomaly
is northward and the Leeuwin Current weakens.

The satellite altimeter observations (Clarke and Li 2004) suggest that
the anomalous ENSO meridional flow hugging the shelf edge is fed by
an anomalous zonal flow between about 22°S and 27.5°S. Thus typically
during a La Niña, there is an anomalous onshore flow between 22°S and
27.5°S that strengthens the Leeuwin Current while typically during an
El Niño there is an anomalous offshore flow between 22°S and 27.5°S that
results in a northward flow anomaly along the shelf edge and a weakened
Leeuwin Current.

It is likely that the anomalous shoreward flow has an influence on the
coastal settlement of the lobsters. When the sea level is higher than normal
at the coast, the anomalous onshore flow between 22°S and 27.5°S will
strengthen the onshore flow and likely will transport more lobster larvae
toward the coast and increase the settlement index; when the sea level
is lower than normal at the coast, the anomalous offshore flow between
22°S and 27.5°S will slow the onshore flow and fewer lobster larvae
will be transported toward the coast. This is consistent with Fig. 12.10.
The logarithmic dependence shown in Fig. 12.10 is like the sea level
anomaly/logarithmic index relationships discussed earlier in Sections 12.4.4
and 12.5.2. It is derived similarly (see Clarke and Li 2004).

12.7. BANANA PRAWNS IN THE GULF OF CARPEN

The Gulf of Carpentaria in northern Australia (see Fig. 12.8) is wide,
shallow and tropical. The surrounding land is low and in many places on
the eastern shore the substantial tide covers the beach and floods the tidal
flats. The many rivers around the Gulf enter the sea as estuaries.

The Gulf climate is monsoonal, about 80% of the rain falling in the
southern hemisphere summer from December to March. The rainfall is
highly variable from year to year because of the influence of ENSO. As
noted earlier in Section 2.2.1 and in Chapter 7, when the equatorial warm
pool moves eastward, the deep atmospheric convection follows it and there
tends to be drought in the western equatorial Pacific, including the Gulf of
Carpentaria. The estuarine environment around the Gulf is highly sensitive
to the rainfall and this has profound implications for the valuable banana
prawn fishery.

Adult banana prawns (*Penaeus merguiensis*) spawn in the open sea and
some larvae reach coastal estuaries where they settle out as post larvae.
Following several months in these coastal nurseries, the juveniles move
offshore and join the adults. The prawns spawn more in March/April and

September/October than at other times during the year and the fishing season is usually between March and May. The fishermen do not trawl in the second main spawning period so that the prawn population can recover. Banana prawns are short-lived animals and the annual catch depends almost entirely on the 1-year class recruited during the year.

The catch fluctuates wildly from one year to the next. It does not seem to be related to the number of juveniles in the coastal nurseries but rather to rainfall, especially in the southern part of the Gulf where most of the banana prawns are caught (Vance et al. 1985). The rainfall drastically changes the estuarine environment and causes the prawns to migrate offshore where they can be caught (Staples 1980). Since about 80% of the rainfall occurs in the southern summer before the banana prawn fishing season in March to May, it is not surprising that the banana prawn catch and southern spring and summer rainfall are correlated. As ENSO causes most of the rainfall variability, it should affect the banana prawn catch. Staples (1983) suggested that this should be the case and Love (1987) showed that the November Tahiti minus Darwin Southern Oscillation Index (SOI) should be a good predictor of the banana prawn catch in March to May of the following year. Figure 12.11 shows that the November SOI and Gulf of Carpentaria banana prawn catch are correlated, although the correlation seems to have fallen during the 1990s. Catchpole and Auliciems (1999) found that the Gulf of Carpentaria tiger prawn catch is also influenced by the Southern Oscillation.

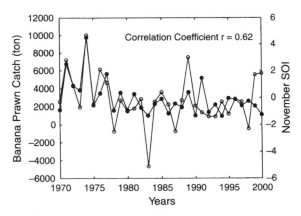

Fig. 12.11 The average November Tahiti minus Darwin Southern Oscillation Index (open circles) and the Gulf of Carpentaria prawn catch for the following year (solid dots), beginning with the 1969 November SOI and the 1970 prawn catch. The prawn catch data were supplied by Malcolm and Bishop (2000) and the SOI data by the Climate Prediction Center of the National Centers for Environmental Prediction, USA (http://www.cpc.ncep.noaa.gov/data/ indices/).

12.8. Green Turtles on the Great Barrier Reef

In 1974 Dr. Colin Limpus became a leader of the turtle research program on Heron Island, Great Barrier Reef (see Fig. 12.8). Based on 5 years of past research in the 1960s it was thought that 200–600 green turtles, *Chelonia mydas*, nest on Heron Island each summer. In the 1974–1975 breeding season Dr. Limpus and colleagues tagged about 1200 nesting female green turtles and local residents said that they had seen more nesting turtles that season than ever before. In the following year the research team, equipped to tag thousands of nesting turtles, encountered only 21 for the entire summer! Some locals suggested that 'Limpus has scared the turtles away.' But we have since learned that huge changes in the number of nesting turtles occur simultaneously for islands over both the southern Great Barrier Reef and also for genetically independent green turtles over the northern Great Barrier Reef. These large-scale simultaneous changes suggest that climate fluctuations may be causing the huge changes seen in the nesting population. To understand how this might happen, we first will need to discuss the biology of green turtles.

Green turtles are large, long-lived marine reptiles. They are herbivorous, grazing on seagrass, algae and mangrove fruit in shallow tropical and temperate coastal waters. In eastern Australia, green turtles breed once every 5–6 years, traveling as much as 2600 km from their home feeding grounds to lay eggs on beaches in the area they were born. Preparation for the breeding migration takes the female green turtles more than a year as they lay down fat reserves and deposit yolk stores in their ovaries.

A likely way that climate changes could influence the nesting population is by decreasing the quantity and/or quality of the green turtle food in their home feeding grounds (Limpus and Nicholls 2000). With less nourishment there would be fewer green turtles able to travel long distances and breed. Since it takes more than a year to get ready for the breeding migration, we might expect that the climate influence on the number of nesting turtles would lead the breeding season by more than a year. Figure 12.12, based on data from Limpus and Nicholls (2000), shows that the logarithm of the number of green turtles nesting is well correlated with the May to October Southern Oscillation Index 2 years before. The sign of the correlation tells us that enhanced nesting occurs about 2 years following major El Niño events and little nesting occurs following major La Niñas. Why should El Niños favor enhanced nesting? If the nutritional hypothesis is correct, then sea grass, algae and mangrove fruit should do better during El Niños than La Niñas, possibly due to a better nutrient supply. Is there

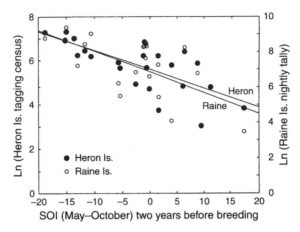

Fig. 12.12 The Tahiti minus Darwin SOI averaged from May to October 2 years before green turtle breeding and the logarithm of the numbers of nesting green turtles at Heron Island (solid dots) and Raine Island (open circles). These islands are in the Great Barrier Reef region (see Fig. 12.8). The data, obtained from Table 1 of Limpus and Nicholls (2000), correspond to the 23 breeding seasons from 1974–1975 to 1997–1998.

in fact a better nutrient supply during El Niños? If so, what is the physical mechanism that causes it? These questions have not yet been answered.

Even though we do not properly understand the mechanism by which ENSO affects green turtle breeding, the turtles' hypersensitivity to ENSO can probably be explained in a similar way to the hypersensitivity of the California zooplankton population in Section 12.4. Specifically, suppose that N is the number of adult female green turtles capable of breeding in a given breeding season and let δN be the change in this number due to a change $\delta(\text{SOI})$ in the May to October SOI 2 years before. Then by similar arguments to the California zooplankton case we have

$$\delta N = -\mu N \, \delta(\text{SOI}) \tag{12.17}$$

where μ is a constant chosen to be positive so that the breeding population is increased ($\delta N > 0$) for a change toward El Niño conditions ($\delta(\text{SOI}) < 0$). Dividing by N and integrating (12.3) gives

$$\ln(N/N_0) = -(\mu)(\text{SOI}) \tag{12.18}$$

where N_0 is the breeding population for neutral conditions ($\text{SOI} = 0$). Equation (12.18) is consistent with the high correlation in Fig. 12.12.

12.9. TUNA AND THE MOVEMENT OF THE EQUATORIAL PACIFIC WARM POOL

About 1300 boats currently fish for tuna in the remote western equatorial Pacific warm pool. The tuna harvest, worth about 1 1/2 billion US dollars per year, is currently about 40% of the world tuna harvest. Most of the tuna caught are skipjack (*Katsuwonis pelamis*).

Lehodey et al. (1997) showed that the skipjack population near the equator (5°S−5°N) migrates east and west along the equator over thousands of kilometers. The zonal migration is well correlated with the zonal displacement of the 29°C isotherm (see Fig. 12.13) which is directly tied to the movement of the eastern edge of the western Pacific warm pool. As discussed in Chapter 2, the displacement of the eastern equatorial edge of the warm pool is fundamental to the coupled air-sea interaction that occurs during El Niño. Thus the tuna migration is strongly tied to ENSO dynamics.

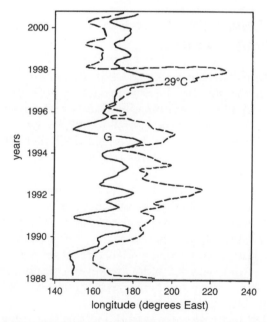

Fig. 12.13 Longitudinal displacement of the 29°C isotherm and longitudinal displacement of the center of the catch per unit effort (CPUE) for shipjack tuna in the equatorial Pacific (5°S–5°N, 120°E–120°W). The CPUE is a measure of the relative abundance of the tuna. Dr. P. Lehodey generously supplied the data.

Lehodey et al. (1997) and Picaut et al. (2001) suggest that the zonal equatorial current convergence near the edge of the warm pool will tend to aggregate plankton, micronekton and hence predators like tuna. This zonal convergence is usually close to the equatorial 28.5°C isotherm, about 1000 km *east* of the 29°C isotherm. But for most of the last decade when measurements are available, the longitudinal center of the skipjack equatorial population is *west* of the 29°C equatorial isotherm (see Fig. 12.13).

Why is the longitudinal center of the equatorial tuna population often to the west of the edge of the warm pool? A major factor not so far considered is that skipjack tuna thrive best in very warm surface water, warmer than the 28.5°C water at the warm pool edge. Therefore, as the warm pool moves eastward during El Niño and westward during La Niña, the skipjack tuna tend to migrate eastward and westward following the warmest warm pool water that lies to the west of the warm pool edge. Lehodey (2001) gives a more detailed discussion of how SST and tuna food both influence the zonal and meridional movement of the skipjack tuna population.

12.10. MIGRATION OF THE BLACK-THROATED BLUE WARBLER

The black-throated blue warbler (*Dendroica caerulescens*) is a songbird which breeds in the summer in the forested regions of eastern North America and spends the winter primarily in the Greater Antilles. Although ENSO originates in the western equatorial Pacific, far away on the opposite side of the world to where these birds live, it can kill them. The population dynamics of these migratory songbirds has been described by Sillett et al. (2000) and what follows is a summary of their work.

Sillett et al.'s results are based on data gathered from 1986 to 1998 at Copse Mountain, near Bethel Town in northwestern Jamaica, West Indies and at Hubbard Brook Experimental Forest, West Thornton, New Hampshire, USA. At both sites there is mature forest relatively undisturbed by humans. The birds have a strong tendency to return to the same breeding site and same overwintering site. Note, however, that none of the observed black-throated blue warblers had both sites in common. Figure 12.14 shows the timing of the bird migration cycle.

Winter in Jamaica is the dry season and the migrant songbirds often have limited food, especially in the late winter just before the spring migration. In El Niño years Jamaica tends to have less rainfall and there are even fewer insects for the warblers to eat. Consequently, just before the

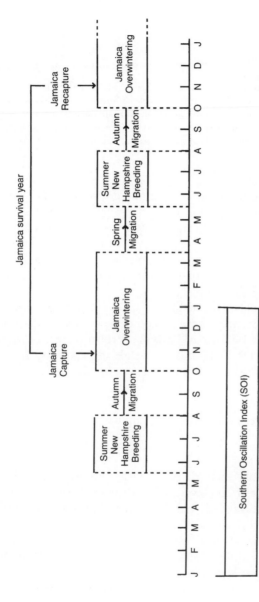

Fig. 12.14 The timing of the black-throated blue warbler breeding, migration and overwintering seasons. The tick marks on the time axis indicate the beginning of the calendar months January (J), February (F), etc. The Southern Oscillation Index (SOI) box shows how the timing of the annually averaged SOI is related to the timing of the capture and recapture of warblers in Jamaica. The latter data were used to estimate annual warbler survival in Fig. 12.15. The Tahiti minus Darwin SOI values were obtained from the US Climate Prediction website www.cpc.ncep.noaa.gov/data/indices/. Comparisons of New Hampshire fledgling weight and fecundity data with the SOI (see text) were done for the New Hampshire breeding season in the SOI year. (This figure is based on data in Sillett et al. (2000).)

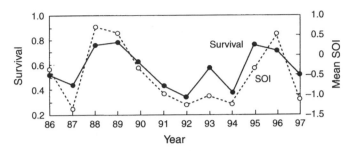

Fig. 12.15 Annual survival estimates of black-throated blue warblers (solid circles) and annually averaged values of the Tahiti minus Darwin Southern Oscillation Index (open circles) for the years 1986–1997. The survival estimates are based on warbler capture and recapture Jamaican data from one year to the next. The capture and recapture timing relative to the annual SOI is given in Fig. 12.14. (Based on data from Sillett et al. (2000).)

arduous spring migration to the summer breeding grounds, many warblers are weak and die on the journey. Opposite effects occur during La Niña years and more warblers survive and return to Jamaica the following year. Figure 12.15 supports this argument – the survival of black-throated blue warblers estimated in Jamaica from one October to the next follows the Southern Oscillation Index which is negative during El Niño and positive during La Niña (see Chapter 2).

El Niño also affects black-throated blue warblers when they breed in New Hampshire. Their main prey there is lepidopteran larvae which tend to be less plentiful during El Niño than La Niña. Probably because of this, there is a strong tendency for fledglings to weigh more during La Niña years than El Niño years. There is also a tendency for the annual warbler fecundity

$$F = (\text{constant}) \times (\text{annual mean no. fledged}) \times (\text{annual mean fledgling mass})$$

$$(12.19)$$

to be higher during La Niña than El Niño years. Consistent with this, the number of black-throated blue warblers returning as yearlings to the New Hampshire site was correlated with both their fecundity (correlation coefficient $r = 0.78$) and with the SOI ($r = 0.59$).

As mentioned earlier, the warblers migrating from the New Hampshire site were not the same ones as those arriving at the Jamaica site. However, ENSO affects whole regions similarly so its influence on fledgling mass and fecundity during the summer breeding season may influence a large percentage of the population, not just the warblers at the New Hampshire site. This seems likely because the number of black-throated blue warblers reaching the Jamaica site in October was positively correlated with both the SOI ($r = 0.71$) and the warbler fecundity F from the preceding New Hampshire summer ($r = 0.63$).

The research described above is a good start on ENSO's influence on a population of migratory birds. There are several unanswered questions. What climatic factors are responsible for the change in the warbler's prey in New Hampshire? What are the physical reasons for these climate changes? Why does El Niño cause drought in Jamaica in winter? Are many other migratory bird populations influenced by El Niño or is the black-throated blue warbler an isolated case?

12.11. CONCLUDING REMARKS

In this chapter I have chosen examples to illustrate the wide variety of ways ENSO can affect bird and animal life. While there are many examples of ENSO's influence, there are also many examples where its influence is negligible. For example, while El Niño affects the herbivorous green turtle on the Great Barrier Reef and also in the Philippines, Malaysia and Thailand (Limpus and Nicholls 2000), it does not affect carnivorous turtles.

The number of reported examples of the influence of ENSO on marine populations has increased greatly since the mid-1990s. Even in those many cases where it is clear ENSO has a strong influence, basic questions have not been answered. There is much work to be done.

REFERENCES

Barber, R. T., and F. P. Chavez, 1986: Ocean variability in relation to living resources during the 1982–83 El Niño. *Nature* (London), **319**, 279–285.

Brink, K. H., D. Halpern, A. Huyer, and R. L. Smith, 1983: The physical environment of the Peruvian upwelling system. *Prog. Oceanogr.*, **12**, 285–305.

Brink, K. H., D. Halpern, and R. L. Smith, 1980: Circulation of the Peruvian upwelling system near 15°S. *J. Geophys. Res.*, **85**, 4036–4048.

Catchpole, A., and A. Auliciems, 1999: Southern Oscillation and the northern Australian prawn catch. *Int. J. Biometeorol.*, **43**, 110–112.

Chelton, D. B., P. A. Bernal, and J. A. McGowan, 1982: Large-scale interannual physical and biological interaction in the California Current. *J. Mar. Res.*, **40**, 1095–1125.

Chelton, D. B., and R. E. Davis, 1982: Monthly mean sea-level variability along the west coast of North America. *J. Phys. Oceanogr.*, **12**, 757–784.

Clarke, A. J., 1991: On the reflection and transmission of low-frequency energy at the irregular western Pacific Ocean boundary. *J. Geophys. Res.*, **96**, 3289–3305.

Clarke, A. J., and M. Dottori, 2007: Planetary wave propagation off California and its effect on zooplankton. *J. Phys. Oceanogr.*, in press.

Clarke, A. J., and A. Lebedev, 1999: Remotely driven decadal and longer changes in the coastal Pacific waters of the Americas. *J. Phys. Oceanogr.*, **29**(4), 828–835.

Clarke, A. J., and J. Li, 2004: El Niño/La Niña shelf edge flow and Australian western rock lobsters. *Geophys. Res. Lett.*, **31**, L11301, doi:10.1029/2003GL018900,2004.

Enfield, D. B., and J. S. Allen, 1980: On the structure and dynamics of monthly mean sea level anomalies along the Pacific coast of North and South America. *J. Phys. Oceanogr.*, **10**, 557–578.

Feng, M., G. Meyers, A. Pearce, and S. Wijffels, 2003: Annual and interannual variations of the Leeuwin Current at 32°S. *J. Geophys. Res.*, **108**(C11), 3355, doi:10.1029/2002JC001763,2003.

Godfrey, J. S., and K. R. Ridgway, 1985: The large-scale environment of the poleward-flowing Leeuwin Current, Western Australia: Longshore steric height gradients, wind stresses and geostrophic flow. *J. Phys. Oceanogr.*, **15**, 481–495.

Jahncke, J., 1998: Las Poblaciones de aves guaneras y sus relaciones con la abundancia de anchoveta y la ocurrencia de eventos el niño en el mar peruano. *Bol. Inst. Mar. Peru*, **17**, 1–13.

Kessler, W. S., 1990: Observations of long Rossby waves in the northern tropical Pacific. *J. Geophys. Res.*, **95**, 5183–5217.

Lehodey, P., 2001: The pelagic ecosystem of the tropical Pacific Ocean: Dynamic spatial modeling and biological consequences of ENSO. *Prog. Oceanogr.*, **49**, 439–468.

Lehodey, P., M. Bertignac, J. Hampton, A. Lewis, and J. Picaut, 1997: El Niño Southern Oscillation and tuna in the western Pacific. *Nature*, **389**, 715–718.

Lenanton, R. C., L. Joll, J. Penn, and K. Jones, 1991: The influence of the Leeuwin Current on coastal fisheries of Western Australia. *J. R. Soc. W. Aust.*, **74**, 101–114.

Lenarz, W. H., D. A. Ventresca, W. M. Graham, F. B. Schwing, and F. Chavez, 1995: Explorations of El Niño events and associated biological population dynamics off central California. *CalCOFI Rep.*, **36**, 106–119.

Li, J., and A. J. Clarke, 2004: Coastline direction, interannual flow, and the strong El Niño currents along Australia's nearly zonal southern coast. *J. Phys. Oceanogr.*, **34**, 2373–2381.

Limberger, D., F. Trillmich, G. L. Kooyman, and P. Majluf, 1983: Reproductive failure of fur seals in Galapagos and Peru in 1982–83. *Trop. Ocean Atmos. Newslett.*, **21**, 16–17.

Limpus, C., and N. Nicholls, 2000: ENSO regulation of Indo-Pacific green turtle populations. In: *Applications of Seasonal Climate Forecasting in Agricultural and Natural Ecosystems. The Australian Experience* (469 pp.), eds. G. L. Hammer, N. Nicholls, and C. Mitchell, Kluwer Academic, Dordrecht, 399–408.

Love, G., 1987: Banana prawns and the Southern Oscillation Index. *Aust. Meteorol. Mag.*, **35**, 47–49.

Malcolm, J. and J. Bishop, 2000: *Northern prawn fishery and Kimberly prawn fishery data summary 2000*. Report LOG-103, Australian Fisheries Management Authority.

Nur, N., and W. J. Sydeman, 1999: Survival, breeding probability and reproductive success in relation to population dynamics of Brandt's Cormorants *Phalacrocorax penicillatus*. *Bird Study*, **46** (suppl.), S92–S103.

Pariwono, J. I., J. A. T. Bye, and G. W. Lennon, 1986: Long-period variations of sea level in Australasia. *Geophys. J. R. Astron. Soc.*, **87**, 43–54.

Pearce, A. F., and B. F. Phillips, 1988: ENSO events, the Leeuwin Current and larval recruitment of the western rock lobster. *J. Cons. Int. Explor. Mer*, **45**, 13–21.

Phillips, B. F., 1986: Prediction of commercial catches of the western rock lobster *Panulirus cygnus*. *Can. J. Fish. Aquat. Sci.*, **43**, 2126–2130.

Picaut, J., M. Ioualalen, T. Delcroix, F. Masia, R. Murtugudde, and J. Vialard, 2001: The oceanic zone of convergence on the eastern edge of the Pacific warm pool: A synthesis of results and implications for El Niño-Southern Oscillation and biogeochemical phenomena. *J. Geophys. Res.*, **106**, 2363–2386.

Pizarro, O., A. J. Clarke, and S. Van Gorder, 2001: El Niño sea level and currents along the South American coast: Comparison of observations with theory. *J. Phys. Oceanogr.*, **31**(7), 1891–1903.

Roemmich, D., and J. McGowan, 1995a: Climatic warming and the decline of zooplankton in the California Current. *Science*, **267**, 1324–1326.

Roemmich, D., and J. McGowan, 1995b: Sampling zooplankton: Correction. *Science*, **268**, 352–353.

Schreiber, R. W., and E. A. Schreiber, 1983: Reproductive failure of marine birds on Christmas Island, Fall 1982. *Trop. Ocean Atmos. Newslett.*, **16**, 10–12.

Sillett, T. S., R. T. Holmes, and T. W. Sherry, 2000: Impacts of a global climate cycle on population dynamics of a migratory songbird. *Science*, **288**, 2040–2042.

Smith, R. L., 1981: A comparison of the structure and variability of the flow field in three coastal upwelling regions: Oregon, Northwest Africa, and Peru. In: *Coastal Upwelling*, ed. F. A. Richards, American Geophysical Union, 107–118.

Smith, R.L., A. Huyer, J. S. Godfrey, and J. A Church, 1991: The Leeuwin Current off Western Australia, 1986–1987. *J. Phys. Oceanogr.*, **21**, 323–345.

Soutar, A., and J. D. Isaacs, 1969: History of fish populations inferred from fish scales in anaerobic sediments off California. *Calif. Coop. Oceanic Fish. Invest. Rep.*, **13**, 63–70.

Soutar, A., and J. D. Isaacs, 1974: Abundance of pelagic fish during the 19th and 20th centuries as recorded in anaerobic sediments off the Californias. *Fish Bull., U.S.*, **72**, 257–273.

Staples, D. J., 1980: Ecology of juvenile and adolescent banana prawns *Penaeus merguiensis*, in a mangrove estuary and adjacent offshore area of the Gulf of Carpentaria. II. Emigration, population structure and growth of juveniles. *Aust. J. Mar. Freshw. Res.*, **31**, 653–665.

Staples, D. J., 1983: *Gulf of Carpentaria Prawns and Rainfall. Proceedings of the Colloquium on the Significance of the Southern Oscillation – El Niño Phenomena and the Need for a Comprehensive Ocean Monitoring System in Australia.* Melbourne: AMSTAC, Department of Science and Technology.

Vance, D. J., D. J. Staples, and J. D. Kerr, 1985: Factors affecting year-to-year variation in the catch of banana prawns (*Penaeus merguiensis*) in the Gulf of Carpentaria, Australia. *J. Cons. Int. Explor. Mer*, **42**, 83–97.

Velez, J., J. Zeballos, and M. Mendez, 1984: Effects of the 1982–83 El Niño on fishes and crustaceans off Peru. *Trop. Ocean Atmos. Newslett.*, **28**, 10–12.

Wolff, M., 1984: Impact of the 1982–83 El Niño on the Peruvian scallop *Argopecten purpuratus*. *Trop. Ocean Atmos. Newslett.*, **28**, 8–9.

Wolff, M., and J. Mendo, 2000: Management of the Peruvian bay scallop (*Argopecten purpuratus*) metapopulation with regard to environmental change. *Aquat. Conserv.: Mar. Freshw. Ecosyst.* **10**, 117–126.

APPENDIX A

EMPIRICAL ORTHOGONAL FUNCTION ANALYSIS (PRINCIPAL COMPONENT ANALYSIS)

Contents

A.1. BASIC IDEA

Empirical orthogonal function (EOF) analysis or principal component analysis (PCA) is a technique used for describing large data sets efficiently. The technique was originally described by Pearson (1902) and Hotelling (1935). It was first used in meteorology by Lorenz (1956). In meteorology and oceanography EOF analysis is used to describe a large number of time series efficiently.

The basic idea is as follows. Consider m time series $X_i(t)$ ($i = 1, 2, \ldots, m$), each time series having zero mean. In order to describe the variability of these time series as simply as possible, we choose that linear combination of the m time series that explains as much as possible of the variability of the m time series. Mathematically, we choose weighting factors p_i so that we maximize the variance of the linear combination $\sum_{i=1}^{m} p_i X_i(t)$; i.e., we choose the p_i to maximize

$$\sum_t \left(\sum_{i=1}^{m} p_i X_i(t) \right)^2$$

where $\sum_t$ denotes summation over all times t of the time series. Notice that the maximum of the above quantity is always found by letting $p_i \to \infty$, so to make the problem well-defined we normalize the p_i by requiring

$$\sum_{i=1}^{m} (p_i^2) = 1. \tag{A.1}$$

In practice the time series $X_i(t)$, $i = 1, \dots, m$ often correspond to time series of a physical variable at spatial locations $j = 1, \dots, n$. The weighting vector $\mathbf{p}$ thus describes a spatial 'pattern' of weighting. It is for this reason that we choose the notation $\mathbf{p}$ for this weighting vector.

The constrained maximization described above is equivalent to the maximization of

$$\varepsilon = \sum_t \left(\sum_{i=1}^{m} p_i X_i \right)^2 + \lambda \left(1 - \sum_{i=1}^{m} (p_i)^2 \right) \tag{A.2}$$

where λ is a 'Lagrange multiplier.' The maximum must satisfy $\partial \varepsilon / \partial \lambda = 0$ and this implies the satisfaction of the condition (A.1). The maximum must also satisfy $\partial \varepsilon / \partial p_k = 0$ and this gives

$$2 \sum_t \left(\sum_{i=1}^{m} p_i X_i \right) X_k = 2 \lambda p_k$$

i.e., in vector form,

$$S_{XX} \, \mathbf{p} = \lambda \mathbf{p} \tag{A.3}$$

where S_{XX} is a matrix having the value $\sum_t X_k(t) X_i(t)$ in the kth row and ith column. Notice that in (A.2) the variability

$$\sum_t \left(\sum_{i=1}^{m} p_i X_i\right)^2 = \sum_t \left(\sum_{i=1}^{m} p_i X_i\right)\left(\sum_{j=1}^{m} p_j X_j\right) = \sum_{i=1}^{m}\sum_{j=1}^{m} p_i (S_{XX})_{ij}\, p_j = \mathbf{p}^{\mathrm{T}} S_{XX}\mathbf{p}$$

$$(A.4)$$

where $\mathbf{p}^{\mathrm{T}}$ is the transpose of the vector $\mathbf{p}$. It then follows from (A.4), (A.3) and the normalization condition (A.1) that

$$\sum_t \left(\sum_{i=1}^{m} p_i X_i\right)^2 = \lambda.$$

$$(A.5)$$

Since S_{XX} is a real symmetric matrix, all its eigenvalues are real and there exist real eigenvectors $\mathbf{p}$ such that (A.1) is satisfied. As the p_i and X_i are real, by (A.5) λ is non-negative and hence all eigenvalues of (A.3) are non-negative. Furthermore, the linear combination of the time series that maximizes the variability is associated with the eigenvector $\mathbf{p}$ corresponding to the maximum eigenvalue ($\lambda_1 \geq \lambda_2 \geq \ldots \lambda_m$). This eigenvector $\mathbf{p}^{(1)}$ is known as the first empirical orthogonal function (EOF) and the linear combination $\alpha_1(t) = \mathbf{p}^{(1)} \cdot \mathbf{X}$ as the first principal component.

A.2. THE HIGHER ORDER EOFs AND THEIR PRINCIPAL COMPONENTS

The eigenvalue problem (A.3) gives not only the first EOF but also the second, third, etc. EOFs associated with the second, third, etc. biggest eigenvalues. What is the meaning of these higher EOFs? What are their properties? Some results for the EOFs and their principal components are given below.

A.2.1. The EOFs are orthogonal

One of the properties of the EOFs is that they are orthogonal. This can be shown by first multiplying (A.3) for the jth eigenvalue by the transpose of the ith EOF to get

$$(\mathbf{p}^{(i)})^{\mathrm{T}} S_{XX}\mathbf{p}^{(j)} = \lambda_j (\mathbf{p}^{(i)})^{\mathrm{T}}\mathbf{p}^{(j)}.$$

$$(A.6)$$

Upon subtracting $(\mathbf{p}^{(j)})^{\mathrm{T}}$ times (A.3) for the ith eigenvalue we have

$$(\mathbf{p}^{(i)})^{\mathrm{T}} S_{XX}\mathbf{p}^{(j)} - (\mathbf{p}^{(j)})^{\mathrm{T}} S_{XX}\mathbf{p}^{(i)} = \lambda_j (\mathbf{p}^{(i)})^{\mathrm{T}}\mathbf{p}^{(j)} - \lambda_i (\mathbf{p}^{(j)})^{\mathrm{T}}\mathbf{p}^{(i)}.$$

$$(A.7)$$

But since $(\mathbf{p}^{(j)})^\mathrm{T} S_{XX} \mathbf{p}^{(i)}$ is a 1×1 matrix it is equal to its transpose and so

$$(\mathbf{p}^{(j)})^\mathrm{T} S_{XX} \mathbf{p}^{(i)} = (\mathbf{p}^{(i)})^\mathrm{T} (S_{XX})^\mathrm{T} \mathbf{p}^{(j)}. \tag{A.8}$$

Substitution of this result into (A.7) and using the fact that S_{XX} is a symmetric matrix and that $(\mathbf{p}^{(i)})^\mathrm{T}\mathbf{p}^{(j)} = \mathbf{p}^{(i)} \cdot \mathbf{p}^{(j)} = (\mathbf{p}^{(j)})^\mathrm{T} \, \mathbf{p}^{(i)}$ gives

$$0 = (\lambda_j - \lambda_i)\mathbf{p}^{(i)} \cdot \mathbf{p}^{(j)}. \tag{A.9}$$

Thus if $\lambda_j \neq \lambda_i$ the EOFs are orthogonal. If $j \neq i$ and $\lambda_j = \lambda_i$, it is still possible to find orthogonal EOFs by the Gram–Schmidt orthogonalization process (see, e.g., Bronson 1989).

A.2.2. The principal components are uncorrelated

Another important property of EOF analysis is that the principal components

$$\alpha_j(t) = \mathbf{p}^{(j)} \cdot \mathbf{X} \tag{A.10}$$

are uncorrelated, i.e., if $i \neq j$ then

$$\sum_t \alpha_i(t)\alpha_j(t) = 0. \tag{A.11}$$

This follows by first noting that the left-hand side of (A.11) is

$$\sum_t \left(\sum_{k=1}^m p_k^{(i)} X_k \right) \left(\sum_{l=1}^m p_l^{(j)} X_l \right) = \sum_{k=1}^m \sum_{l=1}^m p_k^{(i)} (S_{XX})_{kl}\, p_l^{(j)} = (\mathbf{p}^{(i)})^\mathrm{T} S_{XX} \mathbf{p}^{(j)}. \tag{A.12}$$

Then from (A.3) and the orthogonality of the EOFs we have the zero correlation result (A.11).

A.2.3. Representation of $\mathbf{X}(t)$ in terms of EOFs

The m mutually perpendicular EOF vectors $\mathbf{p}^{(i)}$ span m-dimensional space, so for some coefficients α_i we may write

$$\mathbf{X} = \sum_{i=1}^m \alpha_i(t)\mathbf{p}^{(i)}. \tag{A.13}$$

The $\alpha_j(t)$ are actually the principal components defined in (A.10). This can be seen by taking the dot product of (A.13) with $\mathbf{p}^{(j)}$ and using the orthogonality of the EOFs. Equation (A.13) reduces to

$$\mathbf{p}^{(j)} \cdot \mathbf{X} = \alpha_j(t)\mathbf{p}^{(j)} \cdot \mathbf{p}^{(j)}. \tag{A.14}$$

This is equivalent to (A.10) since by definition $\mathbf{p}^{(j)} \cdot \mathbf{p}^{(j)} = 1$ (see (A.1)).

A.2.4. Fraction of variance of the time series explained by the ith principal component

It follows from (A.13), the orthogonality of the EOFs, (A.1), (A.10) and (A.5) that

$$\sum_t \mathbf{X} \cdot \mathbf{X} = \sum_t \sum_{j=1}^{m} (\alpha_j)^2 = \sum_t \sum_{j=1}^{m} (\mathbf{X} \cdot \mathbf{p}^{(j)})^2 = \sum_{j=1}^{m} \lambda_j. \tag{A.15}$$

Thus by (A.5) and (A.15), $\lambda_i / \sum_{j=1}^{m} \lambda_j$ is the fraction of the variance of the time series explained by the ith principal component.

A.2.5. The meaning of the higher order EOFs

Suppose we add an extra constraint to the maximization discussed in Section A.1. Specifically, suppose we maximize $\sum_t (\mathbf{p} \cdot \mathbf{X})^2$ subject to $\mathbf{p} \cdot \mathbf{p} = 1$ *and* $\mathbf{p}^{(1)} \cdot \mathbf{p} = 0$. By orthogonality of the EOFs, any EOF $\mathbf{p}^{(j)} \neq \mathbf{p}^{(1)}$ will find an extremum of $\sum_t (\mathbf{p} \cdot \mathbf{X})^2$ subject to $\mathbf{p} \cdot \mathbf{p} = 1$ and $\mathbf{p}^{(1)} \cdot \mathbf{p} = 0$. By (A.5) these extrema are the eigenvalues $\lambda_2, \lambda_3, \ldots, \lambda_m$. Since by definition $\lambda_2 \geq \lambda_3 \geq \ldots, \lambda_m$, λ_2 is the maximum and hence $\mathbf{p}^{(2)}$ is that vector which maximizes $\sum_t (\mathbf{p} \cdot \mathbf{X})^2$ subject to $\mathbf{p} \cdot \mathbf{p} = 1$ and $\mathbf{p}^{(1)} \cdot \mathbf{p} = 0$.

From the maximization in Section A.1 and the above analysis, we conclude that $\mathbf{p}^{(1)}$ is the vector that maximizes $\sum_t (\mathbf{p} \cdot \mathbf{X})^2$ subject to $\mathbf{p} \cdot \mathbf{p} = 1$ and $\mathbf{p}^{(2)}$ is the vector that maximizes $\sum_t (\mathbf{p} \cdot \mathbf{X})^2$ subject to $\mathbf{p} \cdot \mathbf{p} = 1$ and $\mathbf{p}^{(1)} \cdot \mathbf{p} = 0$. By a similar argument the third EOF $\mathbf{p}^{(3)}$ is that vector which maximizes $\sum_t (\mathbf{p} \cdot \mathbf{X})^2$ subject to $\mathbf{p} \cdot \mathbf{p} = 1$, $\mathbf{p}^{(1)} \cdot \mathbf{p} = 0$ and $\mathbf{p}^{(2)} \cdot \mathbf{p} = 0$. The interpretation of the higher order EOFs continues in this way.

▶ A.3. Physics and EOFs

EOFs are particularly useful when the time series $X_i(t)$ are influenced by a single large-scale process. For example, suppose we have records of very low-frequency sea level along an ocean boundary and that there is no significant alongshore wind. Since the flow is quasi-geostrophic and there is no flow into the boundary, we expect that sea level should be spatially constant along the boundary and have the same time variability (see, e.g., Clarke 1992). This can be described efficiently by (A.13) with one dominant EOF $\mathbf{p}^{(1)}$ with all elements equal.

Since the EOF pattern vectors are orthogonal in space (Section A.2.1) and the principal components are uncorrelated in time (Section A.2.2), EOF analysis is very useful when there are two or more processes with these properties. For example, the low-frequency equatorial thermocline variability is dominated by two such modes (see Sections 5.5 and 5.6 of this book and the original analysis of Meinen and McPhaden (2000)). Specifically, the EOF of the heat content mode due to the recharge and discharge near the equator is of one sign across the Pacific and so tends to be orthogonal with the 'tilt' EOF mode due to the zonal equatorial wind. In addition, based on the physics discussed in Section 7.3, $\partial/\partial t$ of one principal component is approximately proportional to the other. Therefore the correlation is proportional to

$$\lim_{T \to \infty} \frac{1}{T} \int_0^T \phi \frac{\partial \phi}{\partial t} dt = \lim_{T \to \infty} \left[\frac{1}{2} (\phi(t))^2 \right]_0^T \bigg/ T \qquad (A.16)$$

where we have used

$$\frac{\partial}{\partial t} \left(\frac{1}{2} \phi^2 \right) = \phi \frac{\partial \phi}{\partial t}.$$

Since ϕ is bounded, the right-hand side of (A.16) is zero. Thus the two major processes governing equatorial thermocline variability are orthogonal in space and time. It is therefore no wonder that EOFs are useful in the analysis of interannual equatorial thermocline variability.

EOFs are less useful in describing the basic physics when two or more large-scale processes operate and are correlated. Since the principal components are uncorrelated in time, EOFs usually do not clearly separate the processes, making a physical interpretation of the EOFs difficult. But even when the EOF analysis does not clearly show the physics, it can still be very useful. For example, because the basis of EOF analysis is maximization of the variance explained, the 'signal' part of data sets with huge numbers of time series can often be represented by a relatively small number of EOFs and their corresponding principal components. Mathematically, the representation (A.13) may be truncated to a finite number of modes $M \ll m$. This is done, for example, in the practical implementation of the Kalman Filter by Cane et al. (1996) (see Section 10.3).

EOF examples and further theory for EOFs are given, e.g., in von Storch and Zwiers (1999) and Wilks (1995).

▶ REFERENCES

Bronson, R., 1989: *Schaum's Outline of Theory and Problems of Matrix Operations.* McGraw-Hill, New York, 230 pp.

Cane, M. A., A. Kaplan, R. N. Miller, B. Tang, E. C. Hackert, and A. J. Busalacchi, 1996: Mapping tropical Pacific sea level: Data assimilation via a reduced state space Kalman filter. *J. Geophys. Res.*, **101**, 22,599–22,617.

Clarke, A. J., 1992: Low frequency reflections from a nonmeridional eastern ocean boundary and the use of coastal sea level to monitor eastern Pacific equatorial Kelvin waves. *J. Phys. Oceanogr.*, **22**, 163–183.

Hotelling, H., 1935: The most predictable criterion. *J. Educ Psychol.*, **26**, 139–142.

Lorenz, E. N., 1956: *Empirical orthogonal functions and statistical weather prediction.* Technical Report, Statistical Forecast Project Report 1, Dept. of Meteor. MIT, 49 pp.

Meinen, C. S., and M. J. McPhaden, 2000: Observations of warm water volume changes in the equatorial Pacific and their relationship to El Niño and La Niña. *J. Clim.*, **13**, 3551–3559.

Pearson, K., 1902: On lines and planes of closest fit systems of points in space. *Phil. Mag.*, **2**, 559–572.

von Storch, H., and F. W. Zwiers, 1999: *Statistical Analysis in Climate Research.* Cambridge University Press, Cambridge, 484 pp.

Wilks, D. S., 1995: *Statistical Methods in the Atmospheric Sciences.* Academic Press, New York 467 pp.

APPENDIX B CANONICAL CORRELATION ANALYSIS

Contents

B.1. BASIC IDEA

The canonical correlation method finds a linear combination of m predictor time series that maximizes correlation with a linear combination of n predictand time series. Actually, the time series need not be predictors and predictands; the theory applies to maximizing correlation between a linear combination of any appropriate m time series with a linear combination of any appropriate n time series. The mathematical treatment below follows von Storch and Zwiers (1999).

Mathematically, let the m time series be written in the form of a time-dependent column vector $\mathbf{X}(t)$ of length m and the n time series be written as a time-dependent column vector $\mathbf{Y}(t)$ of length n. Then if $\mathbf{p}$ is a time-independent vector of length m and $\mathbf{q}$ a time-independent vector of length n,

$$\alpha(t) = \mathbf{p} \cdot \mathbf{X} \qquad (B.1)$$

and

$$\beta(t) = \mathbf{q} \cdot \mathbf{Y} \qquad (B.2)$$

are linear combinations of the m and n time series, respectively. Our task is to find $\mathbf{p}$ and $\mathbf{q}$ so that we maximize the correlation of $\alpha(t)$ and $\beta(t)$, i.e., we maximize

$$r = \frac{\sum_t \alpha(t)\beta(t)}{\left[\sum_t (\alpha(t))^2 \sum_t (\beta(t))^2\right]^{1/2}} \tag{B.3}$$

where the summation for t is over the length of the time series. Notice that if $\mathbf{p}$ or $\mathbf{q}$ are vectors describing optimal linear combinations and if σ is some scalar then $\sigma\mathbf{p}$ or $\sigma\mathbf{q}$ are also optimal vectors since r in (B.3) remains unchanged. In order to make $\mathbf{p}$ and $\mathbf{q}$ specific, we therefore normalize by requiring that

$$\sum_t (\alpha^2(t)) = 1 \quad \text{and} \quad \sum_t (\beta^2(t)) = 1. \tag{B.4}$$

The problem of maximizing r subject to (B.4) can be written in the Lagrange multiplier form

$$\text{Maximize } \varepsilon = \sum_t \alpha(t)\beta(t) + \zeta\left(\sum_t \alpha^2(t) - 1\right) + \eta\left(\sum_t \beta^2(t) - 1\right). \tag{B.5}$$

Differentiation of ε with respect to Lagrange multipliers ζ or η and setting the result to zero shows that (B.4) is satisfied. Differentiation of ε with respect to any element p_k of $\mathbf{p}$ and setting the result equal to zero gives

$$0 = \sum_t \beta(t)\frac{\partial \alpha}{\partial p_k} + 2\zeta \sum_t \alpha \frac{\partial \alpha}{\partial p_k},$$

i.e.,

$$0 = \sum_t (\mathbf{q} \cdot \mathbf{Y})X_k(t) + 2\zeta \sum_t (\mathbf{p} \cdot \mathbf{X})X_k(t). \tag{B.6}$$

This equation can also be written as

$$(S_{XY})\mathbf{q} + 2\zeta(S_{XX})\mathbf{p} = 0 \tag{B.7}$$

where S_{XY} is the time-independent matrix with kjth element $\sum_t X_k(t)Y_j(t)$ and S_{XX} is the time-independent matrix with kjth element $\sum_t X_k(t)X_j(t)$.

Similarly, by differentiating ε with respect to any element q_k of $\mathbf{q}$ and setting the result equal to zero we deduce

$$(S_{XY})^{\mathrm{T}}\mathbf{p} + 2\eta(S_{YY})\mathbf{q} = 0 \tag{B.8}$$

where S_{YY} is the time-independent matrix with kjth element $\sum_t Y_k(t)Y_j(t)$ and $(S_{XY})^{\mathrm{T}}$ is the transpose of the matrix S_{XY}.

We will assume that the problem has been properly posed so that the matrices S_{XX} and S_{YY} have inverses. Then it follows from (B.7) and (B.8) that

$$-2\zeta\mathbf{p} = (S_{XX})^{-1}(S_{XY})\mathbf{q} \tag{B.9}$$

$$-2\eta\mathbf{q} = (S_{YY})^{-1}(S_{XY})^{\mathrm{T}}\mathbf{p}. \tag{B.10}$$

Multiplying (B.9) by (-2η) and substitution of (B.10) into the result gives the eigenvalue problem

$$4\eta\zeta\mathbf{p} = (S_{XX})^{-1}(S_{XY})(S_{YY})^{-1}(S_{XY})^{\mathrm{T}}\mathbf{p}. \tag{B.11}$$

The vector $\mathbf{p}$ can be determined from (B.11) and the normalization (B.4) which can be written as

$$\sum_t \left(\sum_{i=1}^m p_i X_i(t)\right)\left(\sum_{j=1}^m p_j X_j(t)\right) = 1, \tag{B.12}$$

i.e.,

$$\mathbf{p}^{\mathrm{T}} S_{XX}\mathbf{p} = 1. \tag{B.13}$$

Once $\mathbf{p}$ is known, $\mathbf{q}$ can be determined using (B.10) and the normalization in (B.4) which, analogously to (B.13), can be written as

$$\mathbf{q}^{\mathrm{T}} S_{YY}\mathbf{q} = 1. \tag{B.14}$$

By (B.3) and (B.4), the optimal correlation r is

$$r = \sum_t \alpha(t)\beta(t) \tag{B.15}$$

from which, using (B.1) and (B.2), one can deduce that

$$r = \mathbf{p}^{\mathrm{T}}(S_{XY})\mathbf{q}. \tag{B.16}$$

Since r is a 1×1 symmetric matrix, (B.16) can also be written as

$$r = \mathbf{q}^{\mathrm{T}}(S_{XY})^{\mathrm{T}}\mathbf{p}. \tag{B.17}$$

Multiplying (B.16) and (B.17) together gives

$$r^2 = (\mathbf{p}^{\mathrm{T}}(S_{XY})\mathbf{q})(\mathbf{q}^{\mathrm{T}}(S_{XY})^{\mathrm{T}}\mathbf{p}) \tag{B.18}$$

which upon use of (B.7) and (B.8) gives

$$r^2 = 4\zeta\eta(\mathbf{p}^{\mathrm{T}}(S_{XX})\mathbf{p})(\mathbf{q}^{\mathrm{T}}S_{YY}\mathbf{q}), \tag{B.19}$$

i.e., by (B.13) and (B.14)

$$r^2 = 4\zeta\eta. \tag{B.20}$$

Thus the correlation is the square root of the eigenvalue of (B.11).

Since there is generally more than one non–zero eigenvalue, there is more than one linear combination pair $\mathbf{p}$, $\mathbf{q}$ that results an extremum for ε in (B.5). Canonical correlation eigenvalues and eigenvectors are ordered by decreasing eigenvalues, so, by (B.20), $\mathbf{p}$ and $\mathbf{q}$ corresponding to the first eigenvalue are those that give the highest squared correlation.

Since there are m time series represented by $\mathbf{X}(t)$, the matrix for the matrix problem in (B.11) has dimension $m \times m$. We can equally well solve (B.9) and (B.10) by eliminating $\mathbf{p}$ instead of $\mathbf{q}$ and the resulting eigenvalue problem

$$4\eta\zeta\mathbf{q} = (S_{YY})^{-1}(S_{XY})^{\mathrm{T}}(S_{XX})^{-1}S_{XY}\mathbf{q} \tag{B.21}$$

involves an $n \times n$ matrix. For computational efficiency, we should solve the problem with the smaller matrix. This eigenvalue problem only gives one of the canonical correlation eigenvectors ($\mathbf{p}$ or $\mathbf{q}$), but the other is easily found from either (B.9) and the normalization (B.13) or (B.10) and the normalization (B.14).

B.2. CANONICAL CORRELATION PAIRS AND THEIR CORRELATION PROPERTIES

Let $\mathbf{p}^{(j)}$ and $\mathbf{q}^{(j)}$ be the jth eigenvector pair and let the corresponding linear combinations be

$$\alpha_j(t) = \mathbf{p}^{(j)} \cdot \mathbf{X} \tag{B.22}$$

and

$$\beta_j(t) = \mathbf{q}^{(j)} \cdot \mathbf{Y}. \tag{B.23}$$

A key property of the $\alpha_j(t)$ and $\beta_j(t)$ is that for $i \neq j$, $\alpha_i(t)$ is not correlated with $\alpha_j(t)$ or $\beta_j(t)$ and that $\beta_i(t)$ is not correlated with $\beta_j(t)$.

To show this, we begin with the result from matrix theory that since S_{XX} is a real symmetric matrix, it may be written in the form

$$S_{XX} = R^T D R \tag{B.24}$$

where D is a diagonal matrix and $R^T R = I = R R^T$. The elements of D down the leading diagonal are the eigenvalues of S_{XX} and the columns of R are the eigenvectors of S_{XX}. The eigenvalues of S_{XX} are all non–negative (see the section of text associated with (A.3)–(A.5)). Define the matrix $S_{XX}^{1/2}$ by

$$S_{XX}^{1/2} = R^T D^{1/2} R \tag{B.25}$$

where $D^{1/2}$ is a diagonal matrix with its leading diagonal having non–negative elements equal to the square roots of the elements of D. As the notation suggests, it follows from (B.24) that

$$S_{XX}^{1/2} S_{XX}^{1/2} = S_{XX}. \tag{B.26}$$

Based on (B.24) we also can show, given the existence of an inverse for S_{XX}, that

$$S_{XX}^{-1} = R^T D^{-1} R \tag{B.27}$$

and define

$$S_{XX}^{-1/2} = R^T D^{-1/2} R \tag{B.28}$$

where D^{-1} and $D^{-1/2}$ are diagonal matrices with elements down the leading diagonal being the reciprocal of those of D and $D^{1/2}$, respectively.

Multiplying (B.11) by $(S_{XX})^{1/2}$ gives

$$4\eta\zeta(S_{XX}^{1/2}\mathbf{p}) = (S_{XX})^{-1/2}(S_{XY})(S_{YY})^{-1}(S_{XY})^T\mathbf{p}, \tag{B.29}$$

i.e.,

$$4\eta\zeta\left(S_{XX}^{1/2}\mathbf{p}\right) = S\left(S_{XX}^{1/2}\mathbf{p}\right) \tag{B.30}$$

where

$$S = (S_{XX})^{-1/2}(S_{XY})(S_{YY})^{-1}(S_{XY})^{\mathrm{T}}(S_{XX})^{-1/2}. \tag{B.31}$$

By taking the transpose of $(S_{YY})^{-1}S_{YY} = I$ and using $I^{\mathrm{T}} = I$ and $(S_{YY})^{\mathrm{T}} = S_{YY}$, we have that

$$S_{YY}[(S_{YY})^{-1}]^{\mathrm{T}} = I \tag{B.32}$$

and hence, by multiplying on the left by $(S_{YY})^{-1}$, that the latter matrix is symmetric. It follows from the definition (see (B.28)) of $(S_{XX})^{-1/2}$ that it is also symmetric and hence that S is symmetric. Since S is also real, from (B.30), $S_{XX}^{1/2}\mathbf{p}$ are the eigenvectors of a real symmetric matrix. They are consequently orthogonal and so

$$\left(S_{XX}^{1/2}\mathbf{p}^{(i)}\right)^{\mathrm{T}}\left(S_{XX}^{1/2}\mathbf{p}^{(j)}\right) = 0 \quad \text{for} \quad i \neq j,$$

i.e.,

$$(\mathbf{p}^{(i)})^{\mathrm{T}}S_{XX}\mathbf{p}^{(j)} = 0. \tag{B.33}$$

But

$$\sum_{t}\alpha_i(t)\alpha_j(t) = \sum_{t}\left(\sum_{k=1}^{m}p_k^{(i)}X_k(t)\right)\left(\sum_{l=1}^{m}p_l^{(j)}X_l(t)\right) \tag{B.34}$$

which, after changing the order of summation, gives

$$\sum_{t}\alpha_i(t)\alpha_j(t) = (\mathbf{p}^{(i)})^{\mathrm{T}}S_{XX}\mathbf{p}^{(j)}. \tag{B.35}$$

Thus, by (B.33) and (B.35), $\alpha_i(t)$ and $\alpha_j(t)$ are uncorrelated for $i \neq j$. By a similar analysis, $\beta_i(t)$ and $\beta_j(t)$ are uncorrelated for $i \neq j$. It follows from these results and (B.4) that

$$\sum_{t}\alpha_i(t)\alpha_j(t) = (\mathbf{p}^{(i)})^{\mathrm{T}}S_{XX}\mathbf{p}^{(j)} = \delta_{ij} \tag{B.36}$$

and

$$\sum_t \beta_i(t)\beta_j(t) = (\mathbf{q}^{(i)})^{\mathrm{T}} S_{YY} \mathbf{q}^{(j)} = \delta_{ij} \qquad (\text{B.37})$$

where δ_{ij} is unity when $i = j$ and zero when $i \neq j$.

To show that $\alpha_i(t)$ and $\beta_j(t)$ are also not correlated for $i \neq j$, we must show that

$$\sum_t \left(\sum_{k=1}^m p_k^{(i)} X_k(t)\right) \left(\sum_{l=1}^n q_l^{(j)} Y_l(t)\right) = 0. \qquad (\text{B.38})$$

By exchanging the order of summation, the left-hand side of (B.38) is $\left(\mathbf{p}^{(i)}\right)^{\mathrm{T}} S_{XY} \mathbf{q}^{(j)}$. But by (B.7),

$$(\mathbf{p}^{(i)})^{\mathrm{T}} S_{XY} \mathbf{q}^{(j)} = (\mathbf{p}^{(i)})^{\mathrm{T}} (-2\zeta S_{XX} \mathbf{p}^{(j)}) \qquad (\text{B.39})$$

which, from (B.36), is zero for $i \neq j$.

B.3. The Meaning of the Higher Order Correlation Pairs

We have seen that when we try to maximize the correlation between $\alpha(t)$ and $\beta(t)$ subject to (B.4) we find several extremal solutions $\mathbf{p}^{(j)}$ and $\mathbf{q}^{(j)}$. These extremal solutions are such that, for $i \neq j$, the correlation of $\alpha_i(t) = \mathbf{p}^{(i)} \cdot \mathbf{X}$ with $\beta_j(t) = \mathbf{q}^{(j)} \cdot \mathbf{Y}$ is zero.

To understand these correlation squared extrema and the $\mathbf{p}^{(j)}$ and $\mathbf{q}^{(j)}$, suppose we try to maximize the correlation between $\alpha(t)$ and $\beta(t)$ but now make this correlation subject to both (B.4) *and* the zero correlation of $\alpha(t)$ and $\alpha_1(t)$. All the pairs $\mathbf{p}^{(j)}$ and $\mathbf{q}^{(j)}$ for $j \geq 2$ satisfy the extremal conditions for this problem. Of these, by the chosen ordering of the eigenvalues, the $j = 2$ extremum is largest and therefore the $j = 2$ correlation pair is the maximizing solution. Similarly, we may interpret the $j = 3$ correlation pair as the solution maximizing the correlation between $\alpha(t)$ and $\beta(t)$ but now subject to (B.4) and the zero correlation of $\alpha(t)$ with $\alpha_1(t)$ and also with $\alpha_2(t)$. Correlation pairs for increasingly higher j satisfy problems with increasing numbers of constraints following the above pattern. Note that the $j = 1$ correlation pair gives the overall maximum squared correlation and satisfies the least constrained problem, viz., that the correlation between (B.1) and (B.2) be maximized subject only to (B.4).

B.4. Prediction $Y(t)$ from $X(t)$ Using CCA

The number of possible canonical modes, i.e., the number of non-zero correlations between $\alpha(t)$ and $\beta(t)$, is equal to the number of non-zero eigenvalues of the $m \times m$ matrix S (see (B.15), (B.20) and (B.30)). Since these same non-zero eigenvalues apply to the $n \times n$ matrix in (B.21), the number (say M) of non-zero eigenvalues must be less than or equal to the minimum of m and n.

Suppose we wish to estimate Y a time Δt into the future given X up to time t_*. We can do this by estimating each component $Y_i(t_* + \Delta t)$ of $Y(t_* + \Delta t)$ as the sum of M terms

$$\hat{Y}_i(t_* + \Delta t) = \sum_{j=1}^{M} \gamma_{ij} \alpha_j(t_*) \tag{B.40}$$

where the $\alpha_j(t_*)$ here are associated with the canonical correlation between $X(t)$ and $Y(t + \Delta t)$ instead of $X(t)$ and $Y(t)$. We choose the coefficients γ_{ij} in (B.40) so that the fit is as good as possible in the least-squares sense by minimizing

$$\sum_t \left[Y_i(t + \Delta t) - \sum_{j=1}^{M} \gamma_{ij} \alpha_j(t) \right]^2$$

where the sum is taken over all times t for which $Y_i(t + \Delta t)$ and $\alpha_j(t)$ are available. Differentiating the above expression with respect to γ_{ik} and using the orthogonality of the α_j [see (B.36)] gives

$$\gamma_{ik} = \sum_t Y_i(t + \Delta t) \alpha_k(t). \tag{B.41}$$

In summary, if we have X up to time t_*, then by canonical correlation analysis (CCA) between $X(t)$ and $Y(t + \Delta t)$ we can estimate γ_{ik} in (B.41) using known data up to time t_*. Since the $p^{(j)}$ are known from the CCA, at the present time t_*, we can also calculate

$$\alpha_j(t_*) = p^{(j)} \cdot X(t_*) \tag{B.42}$$

and hence estimate the prediction $\hat{Y}(t_* + \Delta t)$ in (B.40).

B.5. USE OF EMPIRICAL ORTHOGONAL FUNCTIONS IN CCA

It follows from the theory of empirical orthogonal functions (EOFs) (see (A.13)) that we may write

$$\mathbf{X}(t) = \sum_{i=1}^{m} \hat{a}_i(t)\hat{\mathbf{u}}_i = \sum_{i=1}^{m} a_i(t)\mathbf{u}_i \qquad (B.43)$$

and

$$\mathbf{Y}(t) = \sum_{i=1}^{n} \hat{b}_i(t)\hat{\mathbf{v}}_i = \sum_{i=1}^{n} b_i(t)\mathbf{v}_i. \qquad (B.44)$$

In (B.43) and (B.44), $\hat{\mathbf{u}}_i$ and $\hat{\mathbf{v}}_i$ are the EOFs and $\hat{a}_i(t)$ and $\hat{b}_i(t)$ the principal components corresponding to $\mathbf{X}(t)$ and $\mathbf{Y}(t)$, respectively. We define $a_i(t)$, $\mathbf{u}_i$, $b_i(t)$, $\mathbf{v}_i$ as

$$a_i(t) = \hat{a}_i(t)/(\lambda_i)^{1/2}, \mathbf{u}_i = (\lambda_i)^{1/2}\hat{\mathbf{u}}_i, b_i(t) = \hat{b}_i(t)/(\mu_i)^{1/2}, \mathbf{v}_i = (\mu_i)^{1/2}\hat{\mathbf{v}}_i$$
$$(B.45)$$

where the λ_i and μ_i are the variances of the ith principal components. From (B.45), the orthogonality of the principal components and (A.5)

$$\sum_t a_i(t)a_j(t) = \sum_t b_i(t)b_j(t) = \delta_{ij}. \qquad (B.46)$$

The CCA problem is to find $\mathbf{p}$ and $\mathbf{q}$ such that the correlation between $\mathbf{p} \cdot \mathbf{X}$ and $\mathbf{q} \cdot \mathbf{Y}$ is maximized. In terms of the EOF representation above,

$$\mathbf{p} \cdot \mathbf{X}(t) = \sum_{i=1}^{m} a_i(t)\mathbf{p} \cdot \mathbf{u}_i = \sum_{i=1}^{m} a_i(t)P_i = \mathbf{P} \cdot \mathbf{a}(t) \qquad (B.47)$$

where we have defined $P_i = \mathbf{p} \cdot \mathbf{u}_i$. Similarly,

$$\mathbf{q} \cdot \mathbf{Y}(t) = \sum_{i=1}^{n} b_i(t)\mathbf{q} \cdot \mathbf{v}_i = \sum_{i=1}^{n} b_i(t)Q_i = \mathbf{Q} \cdot \mathbf{b}(t) \qquad (B.48)$$

where $Q_i = \mathbf{q} \cdot \mathbf{v}_i$. Thus, the CCA problem is equivalent to finding $\mathbf{P}$ and $\mathbf{Q}$ such that the correlation between a linear combination $\mathbf{P} \cdot \mathbf{a}(t)$ of the principal components of $\mathbf{X}$ and a linear combination $\mathbf{Q} \cdot \mathbf{b}(t)$ of the principal components of $\mathbf{Y}$ is maximized.

One advantage of the above EOF formulation is that the matrix algebra simplifies since, by (B.46), the matrices corresponding to S_{XX} and S_{YY} are both identity matrices. Thus, for example, it follows from (B.11) and (B.20) that the eigenvalue problem for $\mathbf{P}$ is

$$(S_{ab})(S_{ab})^{\mathrm{T}}\mathbf{P} = r^2\mathbf{P} \tag{B.49}$$

with, by (B.13),

$$\mathbf{P}\cdot\mathbf{P} = 1. \tag{B.50}$$

In (B.49), S_{ab} is analogous to S_{XY}, with $\mathbf{X}(t)$ replaced by $\mathbf{a}(t)$ and $\mathbf{Y}(t)$ replaced by $\mathbf{b}(t)$. Also in (B.49), r is the correlation between $\mathbf{P}\cdot\mathbf{a}$ and $\mathbf{Q}\cdot\mathbf{b}$ and is given by (see (B.16))

$$r = (\mathbf{P})^{\mathrm{T}} S_{ab}\mathbf{Q}. \tag{B.51}$$

B.6. Predicting Y(*t*) Using EOFs and CCA

Under the EOF formulation, we have (see (B.44))

$$\mathbf{Y}(t+\Delta t) = \sum_{i=1}^{n} b_i(t+\Delta t)\mathbf{v}_i \tag{B.52}$$

where the $\mathbf{v}_i$ are known from the EOF decomposition. Analogously to (B.40) and (B.41), we can estimate $b_i(t_* + \Delta t)$ by canonical correlation between the EOF coefficients $\mathbf{b}(t+\Delta t)$ and $\mathbf{a}(t)$ based on the times t when $\mathbf{b}(t+\Delta t)$ and $\mathbf{a}(t)$ are known. Note that in (B.43) and (B.44) the EOF sums are usually truncated so that many of the higher order EOFs, which each contribute small amounts of variability similar to noise, are omitted. This usually considerably shortens the number of elements in the vectors $\mathbf{a}$ and $\mathbf{b}$.

Reference

von Storch, H., and F. W. Zwiers, 1999: *Statistical Analysis in Climate Research*. Cambridge University Press, Cambridge, 484 pp.

INDEX

Adjoint data assimilation, 225–7
Adjoint equations, 226–7
Air temperature ENSO
  response, 190–3, 206–10
  mid-latitude zonally symmetric
    component, 190–1, 208–10
  tropical zonally symmetric
    component, 190–3, 206–7
Alongshore particle
  displacement, 259–60
Analysis error covariance (P_a),
  224, 230
Anticyclones, anomalous, 214–15
Atmospheric ENSO model, 196–206
  equations of motion in logarithmic
    pressure coordinates, 197–9
  first vertical mode
    dominance, 204–206
  governing equations, 199–201
  separation into vertical
    modes, 201–204
Australia, 262–74
  banana prawns in Gulf of
    Carpentaria, 267, 271–2
  coastal ENSO flow effect on western
    Australian salmon, 264–6
  green turtles on Great Barrier
    Reef, 267, 273–4
  Leeuwin Current, 264, 266–71
  rock lobsters off western, 267,
    268–71
  western coastline, 262–4

Banana prawns, 267, 271–2
Baroclinic mode:
  two-layer stratification ocean
    model, 43, 44, 98–9
  validity, 103–104
Barotropic mode:
  continuously stratified ocean
    case, 38

two-layer stratification ocean
  case, 43–4
Bias correction, 219–21
Biennial variability, phase-locking
  and, 181–2
Black-throated blue warbler, 276–9
Bottom friction, 83, 85
Boussinesq approximation, 36
Buoyancy frequency, 36

CalCOFI, 256–7
California:
  coastal waters and El Niño, 257
  Rossby waves and ENSO currents
    off coast, 258–60
  sardine fishery, 254, 256–7
  zooplankton off, 256–62
    population and El Niño, 260–2
California Current, 258–61
Canonical correlation analysis
  (CCA), 238–41, 291–300
  application to ENSO
    prediction, 235, 240–1
  basic idea, 291–4
  canonical correlation pairs, 294–7
  higher order, 297
  empirical orthogonal functions
    in, 240, 299–300
  predicting $\mathbf{Y}(t)$ from $\mathbf{X}(t)$
    using, 298
  predicting $\mathbf{Y}(t)$ using EOFs
    and, 300
Characteristics, method of, 95
Christmas Island, 249–50, 252
Climatology and Persistence
  (CLIPER) forecasting scheme,
  234–6
Coastal Kelvin wave, 70–2, 76–8,
  262–3
  physics of propagation, 71–2

301

Printed and bound by CPI Group (UK) Ltd, Croydon, CR0 4YY

21/10/2024

01777257-0001